T0099200

The Practical Book of
BEEKEEPING

The Practical Book of
BEEKEEPING

A COMPLETE HOW-TO MANUAL ON THE SATISFYING
ART OF KEEPING BEES AND THEIR DAY TO DAY CARE

DAVID CRAMP

LORENZ BOOKS

This edition is published by Lorenz Books,
an imprint of Anness Publishing Ltd
info@anness.com
www.lorenzbooks.com; www.annesspublishing.com

If you like the images in this book and would like to
investigate using them for publishing, promotions
or advertising, please visit our website
www.practicalpictures.com for more information.

A CIP catalogue record for this book is available
from the British Library.

Publisher: Joanna Lorenz
Editorial Director: Helen Sudell
Photography: Robert Pickett
Designer: Sarah Rock
Illustrator: Elizabeth Pepperell
Production Controller: Ben Worley

PUBLISHER'S NOTE
Although the advice and information in this book
are believed to be accurate and true at the time of
going to press, neither the author nor the publisher
can accept any legal responsibility or liability for
any errors or omissions that may have been made
nor for any inaccuracies nor for any loss, harm or
injury that comes about from following instructions
or advice in this book. While every effort has been
made to ensure accuracy when researching this
book, the information on the therapeutic and
cosmetic value of honey is often anecdotal and is
not intended as a substitute for the advice of a
qualified professional. Any use to which the
recommendations, ideas and techniques are put is
at the reader's sole discretion and risk.

ABOUT THE AUTHOR
David Cramp MSc started beekeeping while still
serving in the Royal Air Force. Following his service
career he spent a year at the Bee Resarch Unit at
Cardiff University writing a thesis on Drone
Congregaton Areas before heading off to Southern
Spain with his wife to start a beekeeping enterprise.
During his time in Spain he produced organic honey
and wrote extensively for the beekeeping and small
farming press in the UK and USA. He also gave
talks and courses on beekeeping to groups and
individuals from many parts of Europe.

After 12 years in Spain he moved with his family
to New Zealand to manage a 4000 hive beekeeping
operation specializing in pollination and manuka
honey production.

He is a Fellow of the Linnaean Society of London
and a member of the International Bee Research
Association. He now keeps just a few hives as a
hobby and is carrying out further research on drone
honey bees, while working in the New Zealand
education sector.

Honey should not be given to children
under the age of 1 year. In rare instances,
it may contain spores of the bacteria
Clostridium botulinum. Most older children
and adults can cope with this, but because
a baby's digestive system is immature, the
bacteria can multiply in it and cause
problems. If you are pregnant, breastfeeding
or have a medical condition and are
unsure as to whether you should be having
honey, consult your doctor

Contents

INTRODUCTION

When you start keeping bees, a whole new world will open up; there is a fascination in learning about an industrious community, who through co-operation produce a delicious food.

Beekeeping is a hobby like no other. Beekeepers are stockholders of a completely wild and undomesticated creature, the honey bee. Its importance to the health of the planet and the human food chain is enormous. Over the last few decades, a variety of factors have led to a rapid decline in essential pollinating insects. Honey bees can help redress the balance, but even they are suffering from a dramatic decline in numbers due to causes that are not yet not fully understood. Governments are so worried about this trend that they are investing huge sums of money in researching the problems, and they are anxious to attract new beekeepers to the fold in an effort to improve the situation.

A growing awareness of the problems facing our natural world is just one factor that is encouraging more and more men, women and children to take up this fascinating and productive hobby. Others start keeping bees because they see it as an interesting craft that comes with a delicious by-product, and many simply want to open a window on

the natural world; to reconnect with nature. Someone new to beekeeping will want to know what they have to do to become a beekeeper, what special equipment is needed, and what opportunities for further study and advancement there are in beekeeping.

This book is designed to answer those questions by offering a general introduction to beekeeping, touching on all aspects of the subject. It will take you from buying equipment and bees to siting an apiary, whether in the countryside or in town or even on your roof, from day one through all the processes leading up to your first honey harvest. You will also learn about the essential symbiotic relationship between bees and flowers, and which of those flowers will provide nectar at different times during the year. Other information about beekeeping is also introduced, such as the history of beekeeping, exhibiting at honey shows, joining beekeeping associations, harvesting other hive products, and queen rearing. The book also discusses some of the ongoing scientific research into

Above: Beekeeping is undoubtedly one of the most important and fascinating hobbies that you can take part in.

Above: Many beekeepers prefer to keep their hives on roofs; the bees can then fly upward without disturbing anyone.

Above: The WBC (William Broughton Carr) is one of the most attractive hives, however, owing to its double-walled structure, it is not so user-friendly as some other types.

bee diseases and breeding queen bees. Should you wish to increase your knowledge of the subject, a list for further reading is supplied.

Because beekeeping covers many interests, as well as producing honey, you will learn about bees' communication systems and lifestyles, and you may also gain a better knowledge of the wild and cultivated plants and flowers that they pollinate. You will also be helping the environment, increasing the productivity of gardens, orchards and farms and providing a home for a fascinating colony of insects.

Right: To lower the temperature in the hive, bees fan their wings to waft fresh air around; other bees waft out the warm air, until they have cooled the atmosphere.

BEES AND HUMANKIND

People have always been interested in bees, ever since the discovery that they produced honey, and later, that they were pollinators of most of our food crops. Now, more than ever in our threatened natural world, bees are known to be a vital link in the food chain of many animals, including humans. In recent years, there has been a widespread decline in bees and other pollinating insects, causing general alarm throughout the world about the future of our agricultural industry. You can help to preserve the natural cycle by becoming a beekeeper.

Left: This garden is planted with a profusion of bee-friendly plants. Foraging bees are attracted to the flowers when gathering nectar.

THE ORIGIN OF BEES

The exciting discovery of a fossilized bee, preserved in amber, means we can now be reasonably sure that the earliest honey bees first appeared on Earth more than 100 million years ago.

Scientists believe that bees probably evolved from the kind of wasps with mouth parts that could ingest nectar, and then diverged from their hunting wasp ancestors to become pure vegetarians, developing the ability to transport pollen for use as brood food.

The most distinctive differences between present-day bees and wasps are the special adaptations that aid pollination, such as plumose hairs that can hold pollen, and broadened hind legs which help to transport pollen back to the nest as food for the bee brood.

The oldest known bee fossil

Scientists have recently discovered a 100-million-year-old bee beautifully preserved in amber, which suggests that bees began their evolution a long time before that. This newly discovered bee specimen is at least 35 to 45 million years older than any other known bee fossil.

This tiny but well-preserved bee appears to share features with both bees and wasps, and supports scientists' theories of pollen-dependent bees that evolved from their carnivorous ancestors. More than 100 million years ago, the plant world was dominated by conifers, which spread their pollen on the wind. With the gradual evolution of bees, which had the ability to

Above: Anatomy of the worker bee: the mandibles are used to grip and mold wax; the tongue is used to suck up nectar; the sting is located at the end of the abdomen and the Nasonov gland emits the Nasonov pheromone.

pollinate plants, flowering species came into their own. This early fossil specimen appears to be more bee than wasp, and the crossover features give a good idea of the timeline when bees and wasps were separating on their different evolutionary paths. The bee specimen has been named *Melittosphex burmensis*, denoting the location of its discovery in Burma. It has

Right: A worker bee, with legs laden with pollen, is entering the hive to unload her harvest, before setting out to forage again. The pollen is affixed to stiffened, branched hairs on the hind legs of the bee.

Above: A honey bee about to draw nectar and gather pollen from a plant.

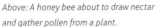

some wasp-like features, such as narrow hind legs, but importantly, it also has plumose body hair and other characteristics of pollinating bees. The fossil bee is in excellent condition, with individual hairs preserved on undamaged portions of its thorax, legs, abdomen and head. The wings are also clearly visible. The fossilized bee is very small; barely 3mm (⅛in) across, but this is consistent with the evidence that early flowers were also very small.

By the time this tiny insect was trapped in its amber prison, bees were already well on their way to serving a completely different biological function from wasps, and from that time on, one of the most remarkable symbiotic relationships ever known began to develop.

Evolution in progress

Flowering plants also began to evolve alongside the essential pollinating bee, and new and larger floral varieties began to appear, which resembled those that exist on our planet now. The discovery of the 100-million-year-

Above: A field of oilseed rape, a crop grown to make oil and animal feed, is attractive to many honey bees. The honey made from rape tends to be extremely abundant, but strong and sweet, so it may require blending with honey of a milder character. It also granulates very quickly, which makes extraction difficult.

old fossil bee coincides with the recent bee genome project and the publication in October 2006 of the genetic blueprint of the honey bee, which reveals surprising links with mammals, including humans.

There are now around 20,000 species of bee, all of which collect pollen to feed their young, and the list of species is constantly growing as bee researchers find new examples.

Because of their huge economic importance to humans, honey bees (*Apis mellifera*) are the best-known bee species today.

Left: These crimson poppies (Papaver orientale) can be grown from seed. They are easily accessible to honey bees and are valuable for their abundant pollen and nectar.

BEES AND THE NATURAL WORLD

For thousands of years, honey bees have played an important role in ecology, by pollinating food crops, helping plants to spread, and providing honey and wax for human use.

Some of the very earliest cave paintings found in Spain, dating from around 7000BC, graphically illustrate honey-hunting expeditions. From these very early times, through the Greek and Roman eras and right up to the present day, humans have always sought out or kept bees. Rock paintings from Africa indicate the importance of bees and honey to early humans, and even today in some tribal cultures, honey is prized as an energy food and it is still the main source of sweetness.

Honey hunting gradually evolved into beekeeping, and archaeologists excavating in northern Israel have discovered evidence of a 3,000-year-old beekeeping industry, including remnants of ancient honeycombs, beeswax and what they believe are the oldest intact beehives ever found.

Above: Bee Farmer, *an engraving by Edmund Evans, 1900 (colour litho) shows both traditional beehives and a skep, which is held by the beekeeper to catch the swarm that is hanging on the branch of a tree.*

Above: A man climbs down a steep cliff face to collect honeycombs from a giant honey bee (Apis dorsata) *nest in Landrung, Nepal. Taking the combs will destroy the nest.*

The honeyguide bird

Over history, humans, birds and animals have joined together to exploit bees and their honey. One of these collaborations features the honeyguide, a bird related to the woodpecker. It feeds primarily on the contents of bee colonies, and is among the few birds that can digest wax. The honeyguide enters populated hives while the bees are torpid in the early morning, or feeds at abandoned hives and those robbed by people or other large animals, notably the ratel or honey badger.

As its name suggests, this bird actively guides people, honey badgers and sometimes baboons, to hives. It attracts attention with wavering,

Above: The ratel or honey badger works with the honeyguide to rob beehives. It is the bee brood that it eats, but in obtaining this, it causes a lot of damage to apiaries.

Above: Bees are vital as pollinators of fruit crops such as lemons.

chattering notes compounded with peeps or pipes. Knowing the location of nests in its area, the honeyguide flies toward an occupied hive or nest and then stops and calls again. It spreads its tail, showing the white spots on each side, and flies up with a bounding movement to a perch, which makes it even more conspicuous.

When human honey hunters reach the bees' nest, they incapacitate the adult bees with smoke and pull the comb apart. After they have taken most of the honey, the honeyguide eats whatever is left.

Bees and pollination

More than one-third of food crops are pollinated by bees. In so doing, they help to increase the yield of produce in orchards, gardens and allotments, and at the same time, ensure the survival of healthy plants. As well as food crops, bees also pollinate wild flowers, which would otherwise have difficulty in setting seed.

The honey bee is uniquely equipped to carry out pollination; grains of pollen stick to the branched hairs on the bee's body and legs, so that when it rubs against the stigma of the next flower,

some of the pollen grains are brushed off, and cross-pollination is achieved. Because honey bees tend to concentrate on foraging from flowers of the same kind, this makes for effective pollination.

Also, since the nectar they collect comes from the same species of flower at any one time, the honey they produce from the nectar has a definite flavour derived from that particular plant.

Above: A greater honeyguide (Indicator indicator) *sits near a honeycomb in Natal, South Africa. The bird makes a chattering sound to indicate the presence of honey.*

HISTORY OF BEEKEEPING

Before the domestication of wild bees in artificial hives, evidence from rock paintings shows that early beekeepers gathered honey from wild bee colonies using smoke to subdue the bees.

Apiculture, or beekeeping, is an ancient practice that goes back thousands of years, and bees have accompanied humans through all stages of civilization.

Ancient Egypt
Research suggests that people in ancient Egypt were the first to develop a culture of beekeeping, from perhaps as early as 4500BC. They kept bees in cylindrical hives made from straw and unbaked clay.

Hieroglyphic carvings have been discovered in temples, on obelisks and on sarcophagi, and show that honey and bees played a significant role in the lives of ancient Egyptians. Royal tombs had wall paintings representing honey, and in most funeral vaults, bees and honey are depicted, while honeycombs, jars of honey and honey cakes have been found in tombs next to the sarcophagi as food for the dead. Most Egyptian medicines contained honey, and it was also offered as a sacrifice to the gods.

Greece
Beekeeping existed in Greece from the earliest times; classical writers refer to it, praising the medicinal and nutritional values of Attic honey, and during the time of Pericles (429BC), the leader of Athens, 20,000 hives were recorded in ancient Attica. Mount Hymettus was abundant with thyme, and honey scented with thyme was daily fare in Athens.

Beekeeping is confirmed by archaeological finds, mythology (the gods were believed to dine on a form of honey called ambrosia), and many texts by classical writers.

Above: Old straw beehives, or skeps, are kept in a stone shelter in a cottage garden in the Museum of Welsh Life, Wales, UK. Inspecting these types of hives and harvesting the honeycomb could not easily be done without destroying many of the bees.

The Roman Empire
Evidence that beekeeping was flourishing comes from the writers Pliny and Virgil. In *Historia Naturalis*, Pliny collected all the wisdom available about honey, and fills several chapters on the subject. Virgil, in the *Georgics*, describes the workings of the hive, and praises honey that is fragrant with the scent of thyme. Guests in ancient Rome were given honey as a mark of hospitality, while snails destined for the royal table were fed and sweetened with honey.

Ancient Britain
Historians believe that the people of ancient Britain used honey for cooking and baking, and to make ale. Penalties and tributes to chieftains were paid with honey and mead. After the Roman conquest, the Britons probably learned much more about apiculture from the invaders, to whom honey was so important that they brought their beehives with them.

United States of America
The first honeybees appear to have been imported into North America in 1622. The Virginia Company in London sent bees and beehives to the state of Virginia on a ship called the *Discovery*. Historical documents claim that the bees multiplied and spread out in North America, but the next import of bees was not until ten years later.

In the 18th century, beekeeping continued to expand, and the bees were farmed to produce honey for domestic use.

Above: Beeswax is used in making good-quality candles that are often used in churches. Candles made with this type of wax burn cleanly with little smoke.

Above: New Zealand Manuka honey, as well as tasting good, has antibacterial and antifungal properties, and is used medically in hospitals, especially in the UK.

SINE CERA (WITHOUT WAX)

Sine cera is the Latin for 'without wax'. Early Roman craftsmen would often use wax to hide any cracks or blemishes that appeared in carved stone statues. The government of the day objected to this deceit and decreed that all such work must in future be sold without wax, or *sine cera*. The word 'sincere' in the English language comes from this ancient Latin root and has come to mean 'genuine' or 'honest'.

Below: Often marble statues had imperfections that would be filled with wax to disguise the flaws.

Use of honey and wax

From the days of ancient Rome, honey has been used in face creams, hand creams, balms, soaps and face masks. Egyptian women ate tablets made from honey and spices to sweeten the breath.

In ancient cultures, the healing properties of honey were recognized and home remedies with honey were deemed to be effective for a wide range of complaints, including respiratory problems and sore throats, kidney and bladder inflammation and gastric ulcers.

Other products of the hive, apart from honey, have made their mark in history. Beeswax candles were used in churches and wealthy households, the cheaper alternative being less bright, smokier and smellier tallow candles.

Above: An ancient apiary, with wooden beehives made from tree trunks and straw, still stands in the countryside of Ukraine.

Early beehives

Once it was realized that bees were as happy in man-made structures as they were in hollow tree trunks, humans provided them with wood or clay pipes, urns, logs and bricks. The Romans and Greeks made pottery hives that resembled large thimbles, and in Europe, from the Middle Ages, wicker skeps or baskets were made in a thimble shape.

UNDERSTANDING BEES

There are already thousands of known species of bee throughout the world, and more species are being discovered all the time. Some bees are among the giants of the insect world, such as the large carpenter bee, which is up to 2.5cm (1in) long, others are so tiny that they are almost invisible; but all have essential roles to play in our natural world. This chapter will guide you through the characteristics of the main species of honey bee and help you to understand their lifestyle and how their co-operative society works.

Left: A swarm of honey bees has collected on a fir tree. The bees surround their queen before a few of them leave to find a new home.

BEE SPECIES

This section describes the honey bee, *Apis mellifera*, and its various subspecies that exist throughout the world. These bees are our main source of honey and beeswax.

Honey bees belong to one of the largest orders of insects, the Hymenoptera, the name of which refers to their heavy wings. This order comprises sawflies, bees, ants and wasps. All live and work in groups in order to survive, that is, they are 'social' to varying degrees.

The 'bee' branch, or Apidae family, comprises Apinae (honey bees), Bombinae (bumblebees) and Meliponinae (stingless bees); each has had a different evolutionary history. Probably the first to become social were the stingless bees.

Out of Africa

Following extensive research probing the origin of the species and the movements of introduced populations, including African 'killer' bees in the New World, scientists now believe that every honey bee alive today had a common ancestor in Africa. The African bees are a hybrid strain of *Apis mellifera* that were bred experimentally and accidentally released, and spread throughout the Americas.

Bees gradually spread into Europe in at least two ancient migrations, resulting in two European populations that are geographically close, but genetically quite different. To explore the movements of bee populations, scientists use simple variations in DNA called SNP (Single Nucleotide Polymorphism) markers. Previous studies relied upon a handful of markers, but now, researchers can use the recently sequenced honey bee genome to locate and compare 1,136 markers, providing a level of detail never before possible.

The genus Apis is composed of ten species, nine of which are confined to Asia. The tenth species, *A. mellifera*, is the exception – it is distributed across sub-Saharan Africa, Central Asia and many parts of northern Europe, and has more than two dozen distinct geographical subspecies.

Bees kept for honey

The honey bees most widely kept commercially and by hobbyists are the subspecies of European honey bees. **European honey bees** (*Apis mellifera*) have been introduced all over the world. They produce a lot of honey and are used to pollinate crops.

Italian bees (*A. mellifera ligustica*) are the most popular subspecies. They are bright yellow or orange, quite gentle in temperament, overwinter well and build up rapidly in the spring. However, they are quick to rob other colonies of their honey and exhaust their own stores in winter, when they will require feeding.

Carniolan bees (*A. mellifera carnica*) originated in central Europe and are grey to brown in colour. They are very gentle, conserve food well in winter and build up quickly in the spring. However, this type of bee is prone to frequent swarming and build comb more slowly than Italian bees.

Northern European black bees (*A. mellifera mellifera*) developed in northern Europe. Most bees from the UK eastward belong to this subspecies – or used to. Since the days of large-scale

Above: Italian bees are clinging to a frame. These bees are a sub-species of Apis mellifera. *Because they are used to a warm climate, they can find it hard to cope with wintry conditions and require more food to compensate.*

Above: The bumblebee (genus Bombus) is a relative of the honey bee; it forms small colonies, eats nectar and feeds pollen to its young. It does make thin watery honey, but in small quantities.

Above: The mining bee (genus Andrena) shown here on tansy, builds its nest on sandy paths, hence its name. These bees are a thinner version of a bumblebee. They feed pollen and nectar to their larvae.

Above: The carpenter bee (Xylocopa violacea) is shown in flight. The bees are so-called because they build a nest in dead wood. Unlike the honey bee, this bee is solitary.

bee movements, which have resulted in hybridization, few honey bees in Northern Europe can now show any form of pure lineage, and the resulting mixes of subspecies can have a variety of temperaments, from mild to fierce. **Iberian bees** (*A. mellifera iberica*) are a European subspecies which can survive in very arid conditions. This Spanish bee is jet black, produces a lot of honey

apparently out of nothing, and attacks beekeepers on sight. For at least 24 hours after any disturbance of the colony, sentries are dispatched to attack anything that may resemble a threat. **Asiatic bees** (*Apis cerana*) is a subspecies that is domesticated in Asian countries including India, Pakistan, Burma and Thailand. They exhibit less swarming activity than wild bees and

tend to live in cavities in human buildings or man-made hives. **Africanized bees** (*Apis mellifera scutellata*) are a subspecies of *Apis mellifera* that were first hybridized in Brazil, then spread through Central and South America. They are very aggressive and are known as killer bees because the accumulative venom of hundreds of them will kill a living creature.

Above: The Africanized bee (A. mellifera scutellata) is a hybrid of African and European stock. It is aggressive, but some beekeepers prefer it because of its resistance to Varroa, and its good honey production.

Above: A honey bee (Apis mellifera) takes nectar from a flower. It sucks it up into its honey crop to be disgorged into a cell in the hive. The bee also picks up pollen, which is stored in the sacs on its legs.

Above: An Asiatic honey bee (Apis cerana) sucks nectar from a camellia. This bee is found in countries of southern and south-eastern Asia, such as India, China and Japan.

AFRICANIZED BEES

New subspecies of honey bee have regularly been introduced to North and South America, including the Africanized bee, a cross-breed of European and African species.

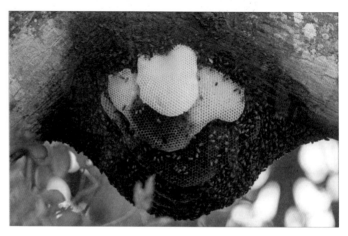

The western and northern European honey bee subspecies *Apis mellifera mellifera* first arrived in North America as early as 1622, following Spanish exploration of the continent. In later centuries, at least eight additional subspecies were introduced from different parts of Europe, the Near East and North Africa.

The Africanized bee, *A. mellifera scutellata*, a subspecies from the savannahs of Tanzania, was introduced to Brazil in 1956. The aim was to breed a strain of bees that would be better adapted to tropical conditions than the European bees, and so would produce more honey. The descendants of these African honey bees rapidly spread northward and southward from Brazil, hybridizing with and displacing previously introduced European honey bees. This produced so-called Africanized 'killer' bees, which can be extremely aggressive. Nevertheless, the bees are now preferred by beekeepers in many

Above left: An Africanized bee (A. mellifera scutellata) gathers nectar from a flower.

areas because of their high productivity – indeed, they are known as producers of 'liquid gold'. Beekeepers use various methods to 'work' these bees, including wearing bee-proof suits rather than ordinary clothes; operating in pairs, one beekeeper smoking the bees while

Above: The beekeeper points to Africanized bees that have colonized a hive of European honey bees.

Above: Layers of comb make up the nest in a tree of Africanized honey bees, or so-called killer bees.

the other manipulates them; and approaching the hives from the rear so as not to alarm the bees. Africanized bees are confined mainly to Africa, North America and South America.

The Africanized queen

The queen stores the sperm in her spermatheca and releases some each time she lays a worker or queen egg, so that the egg is fertilized. If she mates with both European and Africanized drones, some of the resulting queens will be Africanized and some will be Europeans. Africanized queens emerge from their queen cells one day ahead of European queens, and the first to emerge will destroy all the other queens. Thus, the result of a mixed mating will always be a new Africanized queen and, within weeks, a new Africanized bee colony.

Above: An Africanized honey bee swarm is hanging on a tree in Brazil.

Above: A swarm of Africanized bees on a house in Guyana, South America.

Disadvantages of Africanized bees

These bees swarm more frequently and travel farther than other types of honey bee. They are more likely to migrate as part of a seasonal response to lowered food supply. In response to stress, they are more likely to abscond. Compared to other honey bees, they are much more defensive when in a resting swarm. They live more often in ground cavities than the European types, and will guard the hive aggressively, with a larger alarm zone. Unlike other bee subspecies, a higher proportion of guard bees are deployed within the hive.

Advantages of Africanized bees

These bees have a better resistance to Varroa mites and a longer survival rate than other bees. If the scientists can discover what the difference is, they could breed the trait into European bees.

Above: Africanized drones surround a European queen bee on honeycomb in a hive. The queen, centre, is marked with a red dot for identification.

COPING WITH AFRICANIZED BEES

If you live in an area where Africanized bees are found, such as Central and South America or the southern USA, then you should know what to do in the event of an emergency. If you do disturb an Africanized colony, follow these guidelines:

- Run away in a straight line. You should manage to outrun the bees.
- Try and get into a car or building and shut all doors and windows.
- Cover your face and head with a garment or hood. Facial stings, especially in or around the eyes, can be more incapacitating than stings elsewhere.
- Don't try to submerge yourself in a pool or pond. The bees will wait until you surface for air.
- When you are safe, warn the authorities about the location of the colony.

THE HONEY BEE COLONY

The differing functions of bees are a fascinating topic of study, and learning about the way the colony works within the hive is vital to every beekeeper.

During the active beekeeping season, every functional colony will usually hold at least one fertile queen, many thousands of infertile, female worker bees and a few hundred male drone bees. Each of these three castes has very different responsibilities in the hive, and none of the castes could survive independently. It is by learning what each bee is doing, and why, that beekeepers can 'read' the state of their colonies and decide whether they are functioning as they should or whether they have problems.

Above: This bee is raising its abdomen to release a pheromone from its Nasonov gland (scent gland). The pheromone helps worker bees to find their way back to the hive.

The queen bee

To mate and lay fertile eggs are the only tasks of the queen bee. Her lifespan is around four years, but the beekeeper may decide to replace her before she dies. In a populous colony it is often difficult to find the queen, which is why beginners are often advised to start off with a small group of bees in the spring and grow it into a full-sized colony.

A queen may lay up to 2,000 eggs each day in her prime. If it is a worker cell, she will lay a fertilized egg; if it is a drone cell, she will lay an unfertilized egg. She can sting, but will only use her sting to kill off rival queens after she has emerged from her cell. Rarely would she sting the beekeeper.

The worker bee

One of the most interesting and complex insects on earth, the worker bee is an incomplete female, unable to reproduce, but she does just about every other task in the hive. At various stages in her life, she will be a house bee, cleaning and feeding brood; a queen attendant, grooming and feeding the queen; a guard bee, who not only defends the colony but also fans her wings to help with ventilation, and

Above: The queen bee is in the centre of a circle of worker bees, who guard her and tend to her every need, ensuring she is fed on demand and groomed regularly.

forces air over water droplets to cool the hive in hot weather; and finally, a forager, collecting nectar, pollen and propolis. She will also produce wax from exocrine glands under her abdomen, the means to make honeycomb, without which a bee colony cannot survive. The duties she undertakes during her lifespan are regulated pheromonally and are dependent on colony requirements at any one time. Depending on the time of year, a worker usually lives from 15–38 days in the summer, and around 140 days in the winter.

The drone bee

A large, burly bee with a rounded abdomen, the drone is often mistaken for the queen by beginners. He can easily be recognized by his huge eyes, which cover most of his head. He lacks a sting and is designed only for mating. In late spring and summer he will fly to what are known as 'drone congregation areas', which attract virgin queens, and together with other drones he will attempt to mate with the queen in flight. Around 20 drones will be successful in this before the queen departs for home. The successful drones will die immediately after mating as they have everted their internal genitals into the queen and ripped themselves apart. The drone bee has an amazing array of pheromonal sensors, eye facets and strong flight muscles to equip him for this task of sensing and catching a queen in flight.

Bee behaviour in the hive

A colony of bees in the hive shows the role of the different castes. Drone bees fly off on mating flights from around 11 a.m. to 2 p.m. (depending on weather factors). The queen will usually be found on a frame of brood and can be recognized by her long abdomen, which

Above: The worker, or forager, is the most hard-working bee in the hive. Her duties are numerous, and they are driven by whatever is needed in the hive.

is usually a tan colour. She does not like light and she will probably be trying to escape over to the other side of the frame, away from you. Once you've seen her, gently replace that frame in the hive and leave her alone, especially if you have seen eggs in some brood cells – this means that she is laying and all is well. You will see worker bees in their thousands. Some will be returning to the hive with pollen; a few bees will be performing their famous directional 'waggle dance', telling other foragers where to go to gather nectar. Some may be dragging out dead bees; others will be fanning to increase the ventilation in the hive.

Finally, guard bees will be attacking you, but if you have smoked them well, this aspect of life in the hive should not cause you too much of a problem.

Above: The drone is a large bee, with huge eyes. Drones are removed from the hive in the autumn, when there is little food to spare even for productive members of the colony.

THE STRUCTURE OF THE COLONY

A colony consisting of a queen, workers and drones is a dynamic entity that has its own ways of coping with every natural contingency that may arise.

Just as human society organizes itself to carry out many complex tasks such as food provision, defence, household maintenance and rearing its young, so do bee colonies.

Thousands of individuals are effectively programmed so that each bee within the colony knows exactly what it must do at any given time and in any given situation. The colony can house itself, feed itself, defend itself, and if required it can raise another queen and/or move to a better location.

It can even split itself into two separate colonies, one with a new, young queen, by swarming.

Adapting to circumstance

Bees can alter their priorities for the benefit of the colony at any time – for example, if water is required for cooling, more foragers will be sent to collect water, which they carry to the hive in their crop. If many older foragers are wiped out by an accident, for instance by insecticide poisoning, the amount of the worker bee pheromone, ethyl oleate,

in the colony will diminish (older workers have about 30 times as much of this pheromone as younger bees) and younger nurse bees will be able to mature rapidly to fill their places. The presence of this pheromone in the older bees normally prevents younger bees from maturing.

When the queen is fit and well, she produces a pheromone known as queen substance, which draws the worker bees to her and retards their

Above: By wafting their wings, bees can regulate the temperature in the hive and also help to evaporate the moisture from the nectar, which is in cells in the hive, before being transformed into honey.

reproductive growth, thus stopping her rivals becoming queen and laying eggs. The worker bees attracted to the queen will groom her and pass this pheromone around by bee-to-bee grooming contact throughout the hive.

Above: A worker distributes some pollen into the honeycomb cells.

Above: Pollen from various flowers is stored in cells and used for brood food.

Above: Guard bees defend their hive and challenge any unwanted intruder.

Swarming

It is thought that if the hive gets very crowded, there will be less queen pheromone, especially on the outer margins of the nest, and this will signal preparations for a new queen to commence. These preparations begin with making queen cups, which may or may not result in the bees constructing queen cells, raising a new queen and the whole hive swarming.

Decision-making

It is now believed that most decisions that affect the colony are not made solely by the queen, but rather that the older workers make them in response to various internal and external stimuli. For example, they will signal to the colony that it is time to leave the hive in a swarm, and they will indicate to the queen by movement and piping sounds that she must depart on a mating flight or in a swarm. The process by which these decisions are made appears to be controlled by what has been termed the 'anonymous consent' of the colony's workers. Thus, all of these complex tasks require a process of deliberate decision-making involving the whole colony, that is essentially decentralized, and therefore quite different from the strategies that are used by human society.

Above: Worker bees are drawn to the queen by the chemical she emits, and carry out tasks such as grooming and feeding her.

Below: As long as she produces enough pheromone, the queen (centre) will be surrounded by attendants that groom and feed her.

THE LIFE CYCLES OF BEES

The three different castes of bee – worker, drone and queen – have dissimilar life cycles, and the length of their productive life varies too, from a few days to several years.

A successful hive depends on a constant renewal of workers to replace those that have died. It requires cooperation by every type of bee in the nest.

The queen bee

As the only female reproductive unit in the colony, the queen bee is the propagator of the species and without her as the focal point, the colony would dwindle and die out.

The queen lays an egg and deposits it in a cell; she may lay several thousand each day. After a few days, the egg develops into a larva, when it is fed with royal jelly by the worker bees. Once it grows bigger, it begins to pupate; it spins a cocoon and emerges as a young bee.

If the queen dies or for some reason (such as age) stops laying eggs, the bees will raise another queen. The first indication of this is the construction of

cells called 'queen cups' formed on the surface of the comb, hanging downward, although they are sometimes built by the bees just as a precaution.

The old queen lays a fertilized egg in a queen cup, and three days later the egg hatches out inside it. The small larva is fed on royal jelly, a secretion from the hypopharyngeal gland of the nurse worker bees.

After another five days, the bees have built out the queen cup, which now resembles a peanut, and then seal it, leaving the larva inside to pupate and gradually change into an adult bee. The workers may construct several of these queen cells. After another eight days (on day 16 of her life cycle) the new queen will begin to chew her way out of the cell and emerge as an adult. The first queen to emerge will sting through the other queen cells in order

to destroy her rivals, unless prevented from doing so by the worker bees. The workers may decide to retain a queen cell, or they may even allow a second queen to emerge from a cell, in case the original queen does not survive.

A new virgin queen has a difficult journey ahead of her. Within about three days she has to fly far from the hive and navigate to a drone congregation area in order to mate on the wing with drone bees. During this dangerous time out of the hive she may fall prey to a predator, insecticide or foul weather, or she may simply get lost. A reserve queen is always worthwhile. Once a queen has returned from a successful mating flight, the workers will then usually kill the rival queen. Having mated with up to 20 drone bees, the queen will be able to store enough sperm in her spermatheca to fertilize eggs for the rest of her life,

Above: A queen cup in which an egg is nurtured to raise another queen.

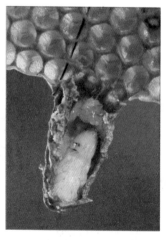

Above: A section through a queen cup reveals the pupa inside.

Above: A new queen in the process of emerging from the queen cup.

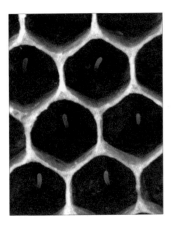

Above: The queen lays an egg in each cell; the eggs resemble grains of rice.

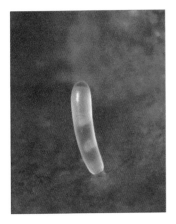

Above: An egg cell, magnified many times to reveal the developing contents.

Above: Larvae hatch from the eggs and are fed with royal jelly, or honey and pollen.

which should be around four years, unless the beekeeper replaces her toward the end of this time. On her return to the hive, she will begin to lay eggs within about 36 hours.

The worker bee

The complex female worker bee will emerge from her cell after 21 days and her lifespan will vary from around three to four weeks during the busy summer months, to up to four months over the winter period, when she will do little or no work.

Above: A close-up of the pupae show how the larvae or grubs have changed.

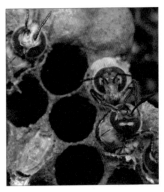

Above: After the pupal stage, the young drone eats his way out of its cell.

The drone bee

A drone takes 24 days to emerge from his cell. He will remain in the hive being fed and looked after by worker bees until he is sexually mature, which takes about 13 days.

He will then take mating flights during the late mornings and early afternoons, whenever it is suitable flying weather, and will either mate with a queen and immediately die, or will remain a member of the colony until removed by worker bees during times of shortage, or at the approach of winter.

Above: A worker emerges by eating her way out of the cell, which is capped with beeswax in order to protect and nurture the pupae.

HONEY AND HONEYDEW

Honey bees make and store honey and honeydew for feeding the colony during the winter months, when flowers are not blooming and no nectar is available.

Unlike other social bees such as bumblebees, honey bees are able to survive periods of shortage because they have developed a method of storing an amount of food in excess of their immediate requirements. This enables a colony to survive as a unit over the winter period, for example, when there is little or no nectar available. Nectar (their energy food) and pollen (their protein food) are both stored, enabling the colony to survive and brood rearing to continue through at least some of the late autumn and during early spring periods.

How honey is made

Honey is derived from nectar, and nectar is essentially an inducement offered by the flower in exchange for bee pollination. Bees are attracted to flowers by their colour, shape and scent, and flowers produce most scent and nectar when the anthers are producing pollen. While the bee is collecting the nectar from the plant's nectaries, it transfers pollen from the anther of one flower to the stigma of another flower, thus allowing the reproductive process to continue. With an increase in the attractiveness and sugar concentration of the nectar, foragers have an increased inducement to remain foraging on one particular flower, and they direct their colleagues to that flower group by means of their symbolic language of movement and dance, described overleaf. Bees will remain faithful to the high-yielding flower so long as it remains a good source of rich nectar.

The nectar is transferred to the bee's honey crop – an enlarged part of the alimentary canal toward the front of

Below: A honey bee sucks up nectar from a flower via its proboscis, or tongue.

Above: A bee returns from foraging with nectar in its stomach, to be made into honey. Pollen grains have collected on the bee as it goes from flower to flower.

Above: The nectar is deposited in cells in the hive after having been treated with enzymes. Moisture is removed by workers fanning the cells before sealing with wax.

Above: Bees cling to a frame which has wax-capped cells containing liquid honey. The bees will be brushed off the frame before the honey is harvested.

the abdomen – and when the bee returns to the hive this is regurgitated to a house bee, who places it into a honeycomb cell. The bee adds enzymes to the nectar, which invert the disaccharide (two-sugar) sucrose into its component sugars, glucose and fructose. This is a very effective process, since it doubles the concentration of sugar molecules packed into any single honey storage cell. Once the nectar is packed into a honeycomb cell, evaporation reduces the liquid to a state in which it will not ferment. The cell is then capped with a light film of wax.

Characteristics of honeydew

Honeydew is a sugar-rich substance excreted by aphids and scale insects as they feed on plant sap. When their mouth part penetrates the phloem (plant tissue that carries nutrients), the sugary, high-pressure liquid is forced out of the terminal opening of the aphid's gut. This excretion forms on leaves and plants and is collected by honey bees, which process it into a dark, strongly flavoured honeydew honey. Although it is similar in looks and viscosity to dark

honey, honeydew contains a broader range of sugars than honey and contains enzymes different from bees'.

This type of 'honey' is much prized in Europe. In times of drought when little or no flower nectar is available, aphid numbers build up substantially and can produce large quantities of honeydew. Ant colonies will often 'farm' aphids, their reward being the sugar-rich sap.

Contraindications to honeydew

There are, however, some disadvantages to the production of honeydew; it can cause problems for bees, because it contains indigestible substances, and has been known to infect bees with a type of dysentery. In some areas, during a cold winter, this has ended in the deaths of many bees. For this reason, in colder climates, beekeepers remove the honeydew before winter sets in. Also, if bees collect honeydew, they will need supplementary feeding, since honeydew does not include pollen, which contains a necessary protein. Pollen is gathered only from flowers rather than aphids.

Right: Aphids, shown here, secrete a sweet liquid called honeydew.

HOW BEES FIND THEIR FOOD

Good siting of the apiary is a help in managing your bees, but the bees need to locate their own food. They then return to the hive with samples, and direct foragers to the nectar source.

The language and the navigational ability of bees are among their most intriguing aspects; bees possess specific symbolic communication skills, enabling them to convey information about food sources to other forager bees. This was the first animal message to be interpreted by humans.

Bees navigate using landmarks, the position of the sun as a fixed object and, if there is no sun, polarized light. They may also take the movement of the Earth into account.

Below: Lavender, alliums, pelargoniums, salvia, thistles and forget-me-nots offer a rich source of nectar for bees.

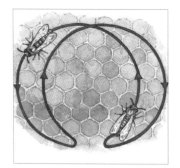

Above: When a foraging bee has found a nearby food source, it communicates with the rest of the hive by doing a 'round dance'. Other forager bees will then look for the food near to the hive, and also by following the scent from the first bee.

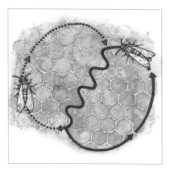

Above: The waggle dance is performed by a forager bee that has found nectar at some distance from the hive. The direction is indicated by the angle from the vertical that the bee describes during its waggle run.

Giving directions

When a scout honey bee finds a good source of nectar, she will fly back to the colony with a sample and, using a language based on movement, sound and the distribution of samples of the nectar, she will describe to other forager bees the distance to the source of the food, and which direction the foragers must fly relative to the sun. The distance is calculated based on the expenditure of energy required to get to the nectar source, so a headwind will cause the scout to increase the representation of distance in the dance. If the bees have to fly over a hill, this energy expenditure will also be taken into account. By handing out samples of nectar to dance watchers, the scout will let them know the odour of the flower, thus aiding their search. Remarkably, this is all done on a

vertical surface in the dark, using the top of the dance or the 'up' position as the notional position of the sun. In other words, the dancer is giving information on a vertical plane about directions that must be followed on a horizontal plane. The dance watchrs project this information on to the horizontal plane when exiting the hive, and fly off in the correct direction relative to the position of the sun.

This mental re-alignment of the directional plane demands an even higher level of language than if the bees danced on a horizontal surface.

The round dance

When a foraging bee finds food close to the beehive, it performs a 'round dance'. The foraging bee runs in a small circle, reversing orientation every one or two circles. This can last for only a few seconds or continue for minutes at a time. The bees watch the dancer and then fly off by themselves looking for the food. They will search in every direction close to the hive while also searching for the particular scent passed to them by the foraging bee.

The waggle dance

If the nectar is farther away from the nest, the bee must give more specific directions, and she will do this by performing the 'waggle dance'.

In a typical waggle dance, the bee runs straight ahead for a short distance, returns in a semicircle to the starting point, again runs through the straight stretch, describes a semicircle in the opposite direction and repeats the whole sequence.

The angle of the straight stretch relative to the top of the hive is the angle that the bees must fly in relation to the position of the sun. During the straight part of the run

Above: In this waggle dance, the bee moves vertically to indicate to fellow foragers that they must fly in the direction of the sun in order to reach the nectar source.

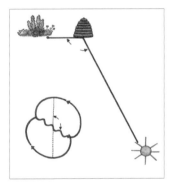

Above: Here the dancer moves at an angle from the vertical, in a downward direction toward the right.

she will waggle or vibrate her body. Now and again, the dance watchers use a squeaking sound to make the dancer pause and give them a sample of nectar to try. The other foragers will then set out and scout around for a while until they have identified the food source by the scent they were given.

Right: A honey bee forages for nectar on Weigela florida 'Variegata'. The shrub attracts pollinating insects, such as honey bees and butterflies.

Above: When the dancer moves at an angle to the left of vertical, it tells the foragers that they must leave the hive and fly at that same angle to the left of the direction of the sun.

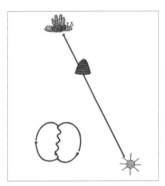

Above: The waggle dance runs vertically downward, indicating that the food is in the opposite direction to the sun.

STARTING BEEKEEPING

Knowledge of beekeeping is not difficult to pick up, and many organizations and local associations are available to help you, especially when you are just starting out. Beekeeping equipment can be bought quite cheaply, and swarms of bees are often free! The more interested you become, the more you will want to experiment with your bees, and the more books and gadgets you will acquire in pursuit of this fascinating hobby. The following pages will explain all you need to know about this absorbing pursuit and how to get everything necessary to start.

Left: A honey bee gathers nectar from a large Cosmos bipinnatus.

CAN YOU KEEP BEES?

If you are thinking of keeping bees, you will naturally begin by assessing two things: how much time you can devote to them, and where you will put the hives.

Although bees are kept in hives, and are to some extent looked after by beekeepers, they are more independent (and perhaps less controllable) than most domestic creatures, and are perfectly capable of fending for themselves for most of the time without the beekeeper's presence. All he or she is doing is housing them in a container acceptable to them, and by way of various manipulations, inducing them to produce honey or other hive products surplus to their own requirements. So, if you are able to visit them every 7–10 days, you will have time to keep bees.

Furthermore, if you are unable to visit them for a few weeks (if, for example, you are away from home), there is no need to worry, so long as you have ensured that appropriately timed Varroa treatments or swarm control measures are undertaken before you go. We will look at these later in the book.

Finding the best site

Bees thrive and produce most surplus honey or other products if the hive is correctly placed, so it is worth investigating your local area and deciding exactly which locations will be most advantageous.

Both rural and urban settings can provide excellent sites for bees, but always remember that, even though bees are very hardy creatures, if placed in dark, dank frost hollows that are prone to flooding they will not thrive, and will probably be unable to produce any surplus honey.

Above: Although bees are very independent, it is necessary to visit them fairly regularly to ensure that they are still alive and thriving.

Any location for bee hives should possess an abundance of flowering plants to provide nectar, and preferably some early sources of pollen such as willow or gorse. Do not assume that the modern countryside is better for bees than urban areas. Vast fields of wheat, in which any small wildflower that pokes its head up is immediately eradicated by the farmer, will not produce much nectar, and thus honey.

Even where there are abundant flowering crops such as sunflowers or borage, it is likely that those flowers will be the single source of nectar that year.

Above: Goat willow (Salix caprea) provides pollen early in the year.

Above: Gorse (Ulex europaeus) flowers in spring, and is a good source of pollen.

Above: Situate your hives at the end of a garden so the bees will be undisturbed.

Guidelines for placing hives

- Sites should have good nectar sources within a range of 2km (1 mile).
- There should be a permanent source of water (not a swimming pool) in the vicinity. Bees like warm water, so a sunny pond with some shade is ideal, if it is quite near to the hives.
- Place the hives in dappled shade, but not directly under trees, and in hot areas, make sure they are away from the danger of forest fires if possible.

- Shelter from wind is important.
- Keep hives out of sight if possible to avoid theft or vandalism, and insure them against loss.
- Ensure that sites have good access, preferably by road.
- Do not site hives in frost hollows.
- Do not site hives under power lines or next to very busy roads.
- If you have several hives, ensure that they are not placed in straight rows all facing the same way, otherwise drifting may occur, when foragers may enter the wrong hive by mistake.

Above: A beehive is sited in an informal garden, which has a pebble-lined pond that is suitable as a water source for the bees.

Above: If there are flowering crops, a field is a good site for several beehives.

SITES FOR BEEHIVES

It can come as a surprise to learn that city sites are very good for honey production, as there is often a richer variety of flowers and trees in urban areas than there are in the countryside.

People in rural or urban areas can keep bees if there is enough food for them at a reasonable distance from the hive.

City versus countryside

Parks, gardens and railway embankments are excellent sources of nectar for bees, as are the chestnut and lime trees that often line city streets. In contrast, the acres of cereal crops in the countryside will not provide useful food for bees. Large cities have a higher temperature than rural areas, which also improves the bees' performance. Bee colonies in cities can produce up to twice the average amount of honey as those in a rural area. The best sites in urban areas are on rooftops and in enclosed gardens.

Keeping bees on rooftops

Many beehives are kept on the roofs of apartment buildings, office blocks or shops. They are out of sight and high up, so they won't annoy neighbours – an all-important factor in city beekeeping.

Above: Even a small garden can accommodate a beehive, and if there are plenty of flowering plants in the area, the bees will produce a lot of honey.

Exclusive bees

One famous London retailer has placed four hives on the roof of its Piccadilly building, from where the bees are able to 'fly high above Mayfair, visiting the grounds, gardens and squares of the best addresses in London, gathering rather superior nectar'.

A hotel in Toronto, opposite the main railway station, has installed three hives on its 13th-floor rooftop terrace to supplement an in-house garden that already provides its nine restaurants with a constant supply of fresh herbs, vegetables and flowers.

Keeping bees in gardens

Another good location for bees is an enclosed private garden. The higher the garden walls, the better, as the bees will be out of sight of potential thieves and vandals. Furthermore, high fences or walls make the bees fly high when leaving the hive, so they will be less of a nuisance to neighbours. Even so, you should keep let your neighbours know when you are about to manipulate your hives, keep them informed about your bees, and finally, give them some of your honey!

Above: The advantage of keeping beehives on a roof is that they are out of sight, and many can fly out of the hive at once without bothering your neighbours.

Above: These impressive hives are on the roof of a famous store in London, UK, called Fortnum and Mason. They produce their own honey to sell in the shop beneath.

Above: Placing hives so that they get the benefit of a high wall in a garden in an exposed rural area will shelter the bees in inclement weather, such as high winds or frost.

WATER SOURCES IN TOWNS

Bees need a good supply of water nearby. Make sure there is a garden pond or a local body of water such as a lake in a park within reach of the bees. If your neighbours have a swimming pool or a pond, consider making a pond in your own garden so that the bees will gather there instead of next door. A garden pond will attract beneficial wildlife to the garden, such as pollinating insects and small mammals that eat pests.

DEALING WITH BEE STINGS

Bee stings do swell and hurt a little and this is normal. However, some people experience a severe allergic reaction, and prompt treatment with epinephrine may be needed.

Many people are apprehensive about getting stung by bees, so be ready to administer first-aid treatment.

Be prepared for stings

Any stings your neighbours suffer, whether they are from bees, wasps or other insects, are likely to be blamed on your bees. Remember that in rare cases, a single bee sting can kill if the person is allergic to bee venom and goes into anaphylactic shock. It is a wise investment to obtain at least third-party insurance in town and city areas where there are many people. There may be existing insurance schemes for beekeepers who are members of local associations, but if not, do make appropriate arrangements.

One of the best ways to reassure your neighbours is to talk to them about your bees. Let them know that even when swarming – in fact, especially when swarming – bees are unlikely to sting anyone unless provoked.

Fear of insects can be a problem. Humans generally have a fear of 'creepy crawlies', which is almost always

Above: Even quite young children can be taught beekeeping. To avoid stings, the child should be appropriately suited and gloved, and should wear a hat and a veil.

because of ignorance about what that insect is likely to do. Education is a powerful weapon in your favour, and it is quite possible that if you talk up the delights of bees and beekeeping, your neighbours may even want to take up the hobby themselves.

All beekeepers will receive stings at one time or another – it is an occupational hazard. Most people find it painful, but the pain usually goes away quite quickly. There is little point in using traditional bee-sting cures, unless you can quickly apply a swift-acting

Above: A magnified sting from a bee. The bee will die after stinging.

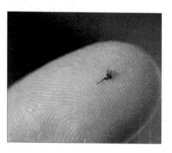

Above: A bee sting, which is still in place in a finger.

Above: Bee stings cause swelling in soft tissues such as those around the eye.

Above: When inspecting the hives and removing the frames that have lots of bees clinging to them, beekeepers should wear gloves, hats and veils to avoid getting stung.

ALLERGY

It is worth protecting yourself against the threat of bee venom allergy by carrying an EpiPen when dealing with your bees. This allergy can be the cause of anaphylactic shock. After being exposed to bee sting venom, the immune system can on rare occasions become sensitized to that allergen and on a later exposure, an allergic reaction may occur.

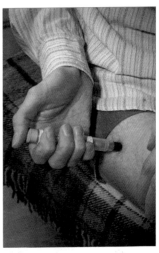

Above: In the event of someone going into anaphylactic shock after a bee sting, it is a wise precaution to have an EpiPen at hand to deal with the symptoms of allergic reaction.

anaesthetic. Obviously there are some areas of the body that will be more sensitive than others, but once you are stung it is just a matter of waiting for the pain to go.

Anaphylactic shock

The risk of anaphylaxis is why many beekeepers, especially those who work alone in the countryside or are in more remote areas, take autoject phials of epinephrine (an EpiPen) with them.

In an emergency, this can be injected into the thigh, which increases blood pressure, opens the airways and saves a life. Anaphylaxis, however, is a rare occurrence, but if it does happen, it is life-threatening.

It is worth remembering that bee sting venom is a very complex compound of substances, and any problems arising from stings should be taken seriously. In commercial bee companies, health and safety regulations

usually require training so that action is taken if problems do occur, and EpiPens must be available at all times.

OTHER PROBLEMS

Bees may also cause spots on washing, cars and house windows during defecation, and this can cause a nuisance, which you must be prepared to remedy, perhaps by resiting your bees or washing the windows.

ESSENTIAL BEEKEEPING EQUIPMENT

Once the hive has been set up and the first bees have entered and made it their home, certain items of equipment are essential to the beekeeper.

You don't need to buy a lot of expensive equipment to start with, but some key items such as a hive tool, a smoker and protective clothing are needed.

The smoker

A good quality, large smoker is a necessity in beekeeping. Cool, dense smoke calms bees. There are various theories as to why this is so, but it seems that the smoke interferes with the sensory mechanism of the bees and induces them to gorge on honey. The effect is almost immediate. Ensure that the bellows of the smoker are sturdy, as these will be the first parts to

Above: A smoker is needed to calm the bees whenever the hive is opened.

wear out, and choose a smoker with a protective grid around it so that you will not pick it up by the metal parts and burn yourself. Buy a large a smoker, as smaller ones do not hold much fuel.

The hive tool

Hive tools come in various shapes and sizes, and are used to prise open hive boxes and frames inside the hive.

Above: When the hive is opened, the bees sense danger and emit certain pheromones. Puffing in smoke disguises the scent, and they remain calm.

They are also used to scrape off propolis and wax from frames, and can be used for hundreds of other tasks around the apiary. Again, it is worthwhile to buy a good quality tool that will last a lifetime.

Protective clothing

Bee suit: A protective bee suit is an essential part of beekeeping, especially for beginners. The suit should cover the body completely, and should have a zip-on hood and veil which can be unzipped

LIGHTING A SMOKER

This can be difficult, especially on a windy day, but if you follow this procedure you should find it straightforward.

1. Shelter from any wind.
2. Prepare the fuel, for example a roll of corrugated cardboard.
3. Place some dry newspaper on the stand in the bottom of the smoker and light it.
4. When it is well lit, push the roll of cardboard into the smoker slowly so that it will also catch fire. Slowly puff the smoker to ensure that air is getting into the bottom of the smoker.
5. As smoke comes out of the top, keep puffing the smoker slightly faster, ensuring that the amount of smoke is increasing.
6. When the roll of cardboard is fully inside the smoker, close the lid and keep puffing until you see cool, dense smoke emerging from the spout. The smoker is now ready for use on the bees.

Above: Children need a protective suit and veil that are zipped together.

Above: A hat and veil with good visibility are necessary for beekeepers.

and pushed back when you are out of range of the bees. Separate hoods and veils that attach by drawstrings are not advised, as the bees seem able to get inside. The hood should have stiffened hoops to keep the veil from the face.

Gloves: Some beekeepers do not like their hands covered; they prefer to have a better 'feel' for what they are doing, but a beginner should wear them, at least initially. Buy a strong pair of kid leather gloves with gauntlets that go up the arm. This helps to prevent bees from entering your sleeve and stinging you, and will also give you sufficient flexibility to be able to handle the frames and boxes as you manipulate the hive. If you want to experiment with greater sensitivity after you have gained some experience, try using a pair of thin washing-up gloves. The bees can sting through them, but you will be able to avoid the worst. Take several pairs to the apiary with you, as they will rip easily on box corners and other sharp edges.

Footwear: This is something many novice beekeepers forget about, but remember that bees are experts at discovering the smallest chink in your clothing, and past masters at the low-level attack. Stings through socks attract more bees to do the same, and from there it is only a matter of time before they start moving up your legs. Many beekeepers wear gumboots with the trousers tucked in to stop bee access.

Above: Gauntlet-style gloves can stop bees getting into sleeves and stinging.

Above: A smoker, showing the bellows that are used to puff out the smoke.

Above: Various hive tools for prising open the box and scraping off wax and propolis.

TRADITIONAL BEEHIVES

The upright beehive with its layered sides and painted walls still has a place in beekeeping, and looks particularly beautiful nestling at the bottom of the garden under some fruit trees.

If you buy new beehives, it is essential that you start off with some that are of a recognized national standard, for which you will easily be able to buy frames, lids, floors, queen excluders and all the other items necessary in beekeeping. Buying non-standard equipment or hives that are not generally used in your area will cause unnecessary problems from the start.

The Langstroth hive

This is the most common hive, used by 75 per cent of the world's beekeepers in the USA, Canada, Australia, South Africa, UK and New Zealand. It was developed in the 1800s by Reverend L.L. Langstroth in the USA, and has set the standard for all modern hives made since then. The most important thing about the Langstroth hive is that it has moveable frames, so the colony is not damaged

Above: The Langstroth hive is not pretty, but it is very practical.

Above: The cottage-style WBC hive is an attractive apiary.

Above: The Dadant hive is an extremely heavy hive used by commercial beekeepers. It is commonly used in Italy, France and Spain.

when the honey is removed. Langstroth realized that bees needed a certain space between combs so that they would not build brace comb between the frames. He spaced the frames in the hive so that when the bees drew out the comb, the 'bee space' of exactly 6.35mm (³/₈in) enabled them to move between the combs at will and ensured that they did not fill this gap with wax or propolis. The same concept is now used in all moveable frame hives, with variations in dimensions to suit local conditions. Wherever you are in the world, you will be able to get standard Langstroth hive parts. However, a word of warning: it does seem that there are as many sizes of Langstroth hive as there are manufacturers. Generally, the parts fit together with a little adjustment, but

some are incompatible, so do check that spare parts for your particular hive are available before you buy.

The Dadant hive

Designed by Charles Dadant, this hive has much in common with the Langstroth and the parts are usually interchangeable. The difference is that the depth of the Dadant frames is increased, thus giving a larger brood box. This has certain advantages, as we will see later. The Dadant is found in many parts of the world and is particularly popular in France and Spain.

Other traditional hives

The WBC hive, which is smaller than a Langstroth, is popular throughout the UK except in Scotland, where

Above: Although it undoubtedly looks good, the WBC, because it has double walls, is harder to use, and is more expensive to buy than the box hives.

the similar-sized Smith hive is more standard. The WBC, invented by William Broughton Carr, accepts standard British National frames, and looks like a proper beehive, rather than a set of packing boxes. The iconic design of the outer covering walls ensures that the hive has retained its popularity among hobbyists in the UK, but the plethora of parts and the gabled lid make it awkward to lift and move, so it is rarely used for commercial beekeeping.

Above: The National is another type of box hive. In this type, the area for the bees to move around is below the frames. This hive is one of the most popular in the UK.

Above: Traditional beehives sited in the Black Forest in Germany.

OTHER TYPES OF BEEHIVE

Beehives come in all shapes and sizes, from old-fashioned and inefficient to excitingly innovative, even including some designed especially for modern urban rooftops.

The traditional beehives described on the previous page, are vertically stacked hives on which new boxes of frames are superimposed on those below, meaning that the hive expands upward. Other hives exist in which the frames are stacked horizontally.

Dartington long hive

This is one such hive found in the UK where although the frames are placed in the hive horizontally, they are standard four-sided frames from which honey can be extracted with the usual extraction equipment. The length of the hive makes up for the lack of additional boxes.

Top bar hives

In third world countries, particularly in Africa, the top bar hive is cheap to produce and easy to handle. This hive has bars to which the honey bees attach and hang wax comb. Unlike the full four-

sided frames used in a Langstroth-type hive or the Dartington long hive, the comb on bars cannot be centrifuged to extract honey and then re-used. This is one of the main disadvantages of the top bar hive, as most extraction equipment is designed to extract honey from standard Langstroth-type frames.

Both the Dartington and the African top bar horizontal hives are easy to operate, and there is less lifting of heavy boxes than with standard vertically stacked hives, making them far easier to use for the elderly or people with disabilities, plus it is cheap to manufacture using basic technology.

The Warre hive

Unlike the African (or Kenya) top bar hive described above, the Warre hive (pronounced War-ray) is a vertical

Left: These hives are made from high-density polystyrene. They have the advantage of being warm and dry, which suits the bees.

Above: The Dartington hive has easily accessible honeycomb, which hangs on bars from the top of the hive.

stacking top bar hive, and as with the Langstroth or Dadant models, more boxes can be added as required. It was designed by the French pastor Emile Warre to be as much like a colony's natural home as possible. The concept of beekeeping in this type of hive also encourages minimal interference with fewer inspections. New boxes are added to the bottom and not the top of the hive which is said to promote the bees' natural tendency to build down, thus ensuring a hive environment that is healthier and better suited to their own needs. Some of these hives have an inspection window in them so the beekeeper can see what is going on in the hive without having to open it up and disturb the bees.

This type of hive is becoming more popular now amongst some hobby beekeepers who want to keep bees as

Don't be put off by some beekeepers who will tell you that modern hives are cruel or unnatural. The bees simply would not remain in them if they were unsuitable. It is true that some commercial moveable frame hives do lend themselves to being moved frequently from crop to crop for pollination, which may well cause stress to bees, but in themselves, modern hives are perfectly satisfactory homes for bee colonies.

Above: This apiary features hives set in a wooden structure. The brightly coloured landing boards help aid bee recognition.

naturally as possible, but it must be remembered (and it is mentioned again later on), that regular inspections for disease should not be minimized even if using one of these hives.

An enterprising company in the UK has manufactured a plastic hive that takes British National standard frames and is specifically designed for small

Below: This is a good example of a Warre hive, complete with inspection windows.

gardens and rooftops. Innovation of this type is a healthy sign, showing that beekeeping is becoming more and more popular.

Polystyrene beehives

High density polystyrene hives were first made in the 1980s. After some initial problems, the makers realised that the material had to be strong enough to resist being chewed by the bees, and heavy enough not to blow down. The advantages of these hives are that they are cheaper, waterproof, and well insulated so that the hives are cooler in summer and warmer in winter. They are also easy to clean, don't rot and do not need painting.

Older hives

Some traditional beekeepers still use the old-fashioned skep beehives made of straw. The bees hang their combs from the inside surface and produce natural round combs. Other hives, such as Spanish cork hives, use a similar principle, where bees hang their natural combs from a pair of crossed sticks placed across the upper part of the hive. While these hives can be fun to experiment with, they are not suitable for frequent inspection or honey extraction, so they can end up being

unattended for long periods of time. Diseases or other problems are liable to build up in them unnoticed, and they then become reservoirs of disease that can then affect neighbourhood bees.

Regular inspections

It is advisable always to use moveable frame hives; they can be easily removed so that efficient inspection can be carried out at appropriate intervals and the beekeeper can check that everything seems healthy in the hive. In these days when the bees are affected by exotic diseases and encounter problems arising from various mites, such as *Varroa*, inspection is a vital part of beekeeping and should be carried out regularly.

Below: The simple design and allowance for natural combs has made the top bar hive popular among natural beekeepers.

HIVE COMPONENTS

Essentially, beehives comprise two types of boxes: a deep box commonly used as the brood chamber, and a shallower box used as a honey box, topped with a lid.

The parts of the hive are basically simple, they are floorless and stackable boxes that hold the frames.

Supers

The shallow boxes, or 'supers', get their name from the fact that they are 'superimposed' or stacked one on top of another, in the upper part of the hive, above the brood box. The supers contain the frames of comb in which bees store their surplus honey. This is the honey that you will harvest. It is possible to use deep boxes as honey supers, but a deep box full of honey is very heavy and difficult to move. It is also easier to distinguish deep brood boxes from shallow honey supers in storage, but it does mean that the boxes are not all interchangeable.

Brood nest

This space (also known as the brood chamber) is a deep box holding 10 frames of comb. It is reserved for the queen and most of the bees of the colony to rear brood and store honey purely for their own use. Either one or two boxes (also known as hive bodies)

can be used for a brood nest. Two hive bodies are common in cold winter regions, whereas beekeepers in areas with mild winters can successfully manage with only one.

Hive stand

This keeps the hive off the ground so that damp cannot rise up and permeate the hive. There should always be a good flow of air underneath the hive. Hive stands can be made of almost anything; old tyres, wooden pallets (preferably rot- and insect-proofed),

Above: A full hive box with crown board and floor, made from the components shown on these pages.

bricks, iron rails, or even purpose-built stands that have been designed especially for the beehive.

Bottom board or floor

The wooden stand on which the hive rests is known as the bottom board. The front is open, allowing access for the

Above: Frames for brood box and super

Above: Brood box with frame

Above: The floor of the hive

bees. Many beekeepers now use stainless steel mesh floors with a wooden surround. The mesh allows for good hive ventilation and ensures that if rain gets into the hive it immediately drains out. Mesh floors also hinder Varroa mites, which fall out of the hive if they are dislodged from a bee and are then unable to return. If you use a solid wooden floor, always ensure that the hive is slightly tilted forward on its stand so that any rain entering the hive will run out of the entrance and not accumulate in the hive. Remember that damp conditions can kill bees.

Queen excluder

This screening device is made from wire or plastic mesh with a gauge fine enough to prevent only the larger queen moving through it. The excluder is placed between the brood nest and the honey supers to keep the queen in the brood nest, so that she will not be able to lay her eggs in the honey supers. An excluder is not usually necessary if two brood boxes are used.

Frames and foundation

These wooden frames hold sheets of beeswax foundation that are imprinted with the shapes of hexagonal cells. Beeswax foundation helps the bees to build straight combs. Hoffman and Manley frames are popular, and are constructed from plastic. More and more

beekeepers are using these because of their ruggedness and long-lasting qualities. Dip these frames in beeswax prior to using them, so that the bees draw them out into cells more swiftly.

Crown board

This is a cover to fit on top of the brood box to keep a constant temperature in this part of the hive. The holes in the centre allow the bees to be fed without the need to remove the crown board. The crown board also prevents bees from attaching comb to the lid, but is often omitted by beekeepers.

The outer cover or lid provides weather protection and is covered with tin or zinc sheeting.

Above: A frame feeder looks like a standard frame, but it is filled with syrup so that the bees can feed from it if necessary. It slots into the hive in the place of one of the frames filled with wax foundation.

Feeders

These hold sugar syrup to be fed to bees when needed. The most convenient types for the new beekeeper are frame feeders, but since they do not hold much liquid, they have to be refilled often.

Hive straps

These straps hold hives together if you change the location of your hives. The toughest straps are of metal.

Above: Queen excluder

Above: A frame with wax foundation

Above: Crown board with bee escapes

HOW A BEEHIVE WORKS

A modern beehive allows the bees to live and work comfortably and safely while the beekeeper removes the hive products, with the least possible disruption to the bees.

We have seen that essentially the beehive is a collection of boxes stacked on top of each other with a floor, an opening and a lid. Now it is time to look at how it all works.

Frames

An important operational feature of the beehive is the frames that contain the foundation on which the bees build (or draw out) the comb (in the brood box) or honeycomb (in the honey supers), in which they store the honey. In a modern hive, these frames are straight so that they can be lifted out by the beekeeper at will.

Frames can be deep or shallow, depending on where they are to be used. In the brood box, specially designed shallow frames are self-spacing and constructed so that it is difficult for the bees to glue them together with propolis. In the honey supers, straight-sided Manley frames allow the bees to draw out honeycomb to its greatest extent for maximum honey storage.

The straight sides facilitate the development of even honeycombs and make uncapping easier. Each frame is separated by its neighbour and by the side of the frame, or sometimes by a plastic device. This separation ensures that bee space between the frames is maintained. Hive frames are usually sold with the hive, and in this case there is no choice in their design. There are usually ten frames in brood boxes and eight or nine in the honey supers. The frames hold pre-prepared wax sheets of foundation that are embossed with hexagonal cell shapes ready for the bees to draw out into brood or honeycomb. The foundation may be wired or unwired; if wired it is stronger and can be reused.

Honey supers

Above the queen excluder, the honey supers, with foundation wax or drawn-out honeycomb, will allow the bees to store surplus honey. The foragers will do this as long as a honey flow continues. This is the honey that you will harvest.

Brood box

This box holds the bees' nest and is a very busy area, since most of the activities in the hive take place in it. The queen lays her eggs here, the larvae are reared in this part of the hive, thousands of baby bees are nurtured here, and there is a storage area for food to sustain the brood.

Two brood boxes may be required for a prolific queen. It works as follows: The queen lays her eggs in this box and is prevented from going up into the honey boxes and laying eggs there by the queen excluder.

The brood is kept at a temperature of 34°C (93°F) – the bees either heat or cool the area themselves to maintain this temperature by wafting their wings.

Brood emerge from their cells here; nurse bees clean the used cells and help to feed the larvae with honey and pollen, which is stored in the brood nest. Every other operational task in the hive is also carried out in this area.

Above: A beekeeper holds a super – a box in which the bees draw out the comb and store their supplies of honey.

Above: These frames of foundation are pre-made. They have wax on each side moulded into hexagon shapes.

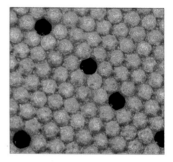

Above: This brood looks compact, with a good covering of wax cappings, and many of the cells are full.

Above: Regular inspection of the hive is essential to check the brood pattern, in which there should be many capped cells, and to check the health of the bees.

Above: Nurse bees in the brood chamber or deep box help to feed the growing larvae and clean out empty cells. Most of the colony live here, so it is hot and very busy.

Nurse bees

For the first few days of a larva's life, the queen and workers both have the potential to reproduce, and the nurse bees, which produce the royal jelly, feed them all the same diet of jelly for 2–3 days. This then changes, and only the larva destined to be queen receives 100 percent royal jelly; the workers are given a mixture of jelly, pollen and honey. Royal jelly stimulates the queen's egg-producing organs. After pupating and emerging as adult bees, the workers, unlike the queen, are unable to reproduce.

Inspecting the hive

The moveable frame hive allows the beekeeper to check the hive at frequent intervals. With more and more disease affecting bees, inspections are the best way to learn about the bees' health and prevent disease spreading to other apiaries.

FORAGER EXCLUDERS

Many beekeepers believe that queen excluders, which prevent her entering the honey supers and laying eggs where the surplus honey is stored, can also inhibit foragers, or workers, from going up into the honey supers to deposit nectar. There are two ways that you can overcome this problem and help to ensure that foragers are able to gain speedier access to the honeycombs.
• You can drill a bee-sized hole in each honey super toward the front of the hive.
• Stagger the supers slightly in fine weather so that the frames of the box below are revealed.

Above: The queen excluder is a physical barrier that prevents the queen entering the honey supers and laying her eggs there. The supers are above the brood chamber, and provide an area where the bees can store surplus honey.

MAKING YOUR OWN HIVE

If you are interested in carpentry, assembling your own hive can be an exciting project, and gives the beekeeper a great sense of achievement once the honey starts flowing.

Decide on the design of your hive early on. This should be the most commonly found moveable frame hive in your area – probably the Langstroth, Dadant or British National design.

The Langstroth is very versatile, because you can get standard parts anywhere. If you want some style in your beekeeping, go for the iconic WBC design, which uses British National inner parts (although Langstroth-size WBCs have been made). The WBC is more complicated to build.

ASSEMBLING THE HIVE

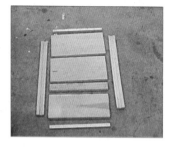

1 Lay out and glue the hive floorboard parts together.

2 Separate out the tongue-and-groove side rails for the floorboard.

3 With the floorboard flush with the ends of the rails, hammer nail a toward each end of each side rail.

4 Check if your floorboard is reversible; it will have a deep side that makes a large summer entrance; flip the board for winter.

5 Nail and glue on the top and bottom cleats on each end to match height of the side rails.

6 The four sides of a single shallow hive box are shown. Note the box joints.

7 Use a wooden stick and apply a dab of glue to all the box joints.

8 Make into a box and nail each bottom corner, using galvanized nails.

Buying a kit

This is the easiest and quickest way, and a typical kit design is shown opposite. You will need a hammer and nails, some good wood glue and some clamps. Most kits have parts with instructions.

Building from plans

This is suitable for a more skilled carpenter. Obtain a box from a local beekeeper or bee supply store to use as an example, with detailed plans from the internet or bee supply store.

Prepare your space, equipment and tools. You will need a hammer and nails or a staple gun, screws, wood glue and clamps. Either buy the wood or have it cut for you, and stack it neatly. Buy tin to cover the lids and cut it to size. Thin aluminium or tin sheeting is the easiest to cut, fold and pin down at the sides, but it is expensive.

Construction materials

Wood is the traditional material for beehives, and is best for the amateur, although expanded polystyrene and plastic hives now exist. Buy good quality pine or red cedar. Cheap wood can be full of knots. As the wood weathers, the knots become knot holes, and you will have more entrances than you want. Avoid using plywood as it is very heavy.

Frames can also be made but are probably better purchased ready-made, especially if you want plastic ones. They can be made up from kits (shown here).

Floors made of stainless steel mesh set in a sturdy frame are popular. Mesh floors assist with ventilation, Varroa control, water drainage and hive cleanliness. They are expensive, but worth the extra cost.

Lids should be sturdy, with covers made of galvanized iron, aluminium, or zinc sheeting. The lids can be telescopic or flat but must fit snugly to the top box. For the novice, telescopic lids are safer.

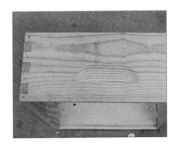

9 *Ensure all is aligned before nailing, and that the assembly remains square.*

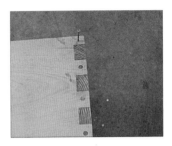

10 *Drive nails that are 3.8cm/1½in long into all the pre-drilled holes.*

11 *The completed hive body or box is roofless and floorless.*

12 *Find the grooved top bar frames – a top bar, two side bars and bottom bars.*

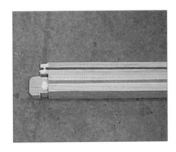

13 *Apply a dab of glue to each end of the top and bottom bars.*

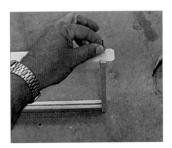

14 *After squaring up the frame, nail the top bar to each of the side bars.*

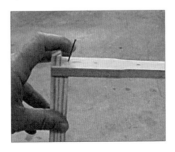

15 *Drive the nail in at an angle at each end, through the end bar into the top bar.*

16 *Fit all the frames that you have made into the hive body.* ⟶

ONE TO AVOID

Many beekeepers and hive builders suggest that castellated frame spacers are a good idea. These spacers are fixed along the revetment from which the frames hang. This seems like a good idea until you start to use them, when they quickly become one of the most irritating things you will ever encounter in your beekeeping career. Just let the frames space themselves, as they will do naturally if you buy Manley or Hoffman frames.

Joining the pieces

Use 50mm (2in) nails or a staple gun with long staples to join the pieces of the hive. Good quality staples work well and speed up the process of assembling the boxes. If you use nails, remember that the holding power of nails driven into the end grain is increased by driving them in at an angle. For additional strength, end pieces may be glued before nailing, and screws can also be added to strengthen the structure.

The hive kit shown here uses plastic foundation, which is more durable than wax foundation, especially when extracting your honey. It is perfectly acceptable to the bees but of course you could not make comb honey from this. Pre-wired wax foundation is also common, and is just as easy to attach to the frames. Simply slot it into the groove in the top frame and pin it in place through the frame. Your bee supply store will advise you further.

Extra features

An advantage of making your own hive is that you can choose the best features and add on extra elements that make your life easier, as well as ensuring that the bees are more comfortable, and eliminate problems. Some of the most common extra features follow.

17 *Press and snap the plastic foundation into place to fit into the frame.*

18 *Check the finished frame is as shown, with the plastic foundation in place.*

19 *Place the frames and foundation into the box hive.*

20 *Assemble the top and side rails, ready to make the inner cover.*

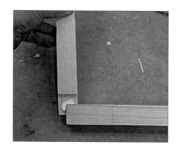

21 *Glue and insert the side rails into the end cleats.*

22 *Slide the top on to the side rails, and fit the other cleat.*

23 *Drive in two nails to add strength to the corners; flip over and nail again.*

24 *Place the inner cover on top of the hive, with a slot for feeding or bee escape.*

WEATHER PROOFING

• To ensure a long, trouble-free life, hive bodies must be treated with wood preservative or paint.
• Apply a good quality exterior latex paint or wood preservative.
• Make sure that none of the preservatives contain insecticide for wood borers (most do).
• Do not paint or stain the inside of the supers.

Alighting board If you are not going to move your hives and stack them together on a truck, you can construct the floors so that they extend out of the front of the hive and provide a landing board, or alighting board, for the bees. This is not really necessary to the bees, but it provides you with an opportunity to watch your bees and study their behaviour, such as guarding and fanning. WBC hives have integral alighting boards.

Above: An alighting board.

Handles Most hives have inset hand holds in the side of the boxes to facilitate lifting, but it is a strain on the fingers if the boxes are heavy. Battens nailed to the front and back of your boxes are better. These may look unwieldy, but they make lifting heavy boxes full of honey much easier. It is simpler to affix battens than it is to make neat holes in the hive body.

Floor lugs These are simply heavy-duty staples or nails at each corner of the floor, a little way in from the edge, around which the brood box sits.

25 *Gather the parts for the outer cover or lid, which slot together.*

27 *Find the end rails for the lid, which have the grooves that the lid board will slot into to ensure they are flush and fitting.*

29 *Glue and nail the top cleats in place; use enough nails to hold them firmly.*

They help the beekeeper to place the box on the floor more accurately, and prevent the box from slipping or getting knocked askew.

Honey super entrances If there is a hole in the front of each super box, in a heavy flow, bees bearing nectar do not have to enter the hive at the bottom and work their way up through the hive. However, they do provide access for hive robbers.

26 *Identify the tongue-and-groove pieces of the outer cover.*

28 *Fit the tongue and groove pieces. Tap them into place until they fit snugly. Glue and nail the end rails to the lid board.*

30 *The lid should be painted and covered in metal sheeting.*

Above: The finished beehive.

ACQUIRING YOUR BEES

There are several ways to obtain your bees, but before you bring them in, check that you are completely prepared for the new inhabitants of the hive.

It is important to have a suitable apiary site ready, and ensure that your hives are fully stocked with comb or foundation as appropriate. Collect all the ancillary equipment together, including adequate clothing, a smoker and a hive tool. You could ask a more experienced beekeeper to be with you on the first day in case of problems, particularly if you are just starting beekeeping.

Buying a colony in a hive

This is the most straightforward method of obtaining bees. Check the following:

- The hive should contain at least one brood box and two honey supers, a queen excluder, and also an entrance reducer.
- The bees in the brood box should cover at least six frames of brood, pollen and honey. Inspect the combs in the deep supers for brood quality. Capped brood is tan to brown in colour and should not have a pepper-pot appearance with numerous empty cells among the brood cells.
- Ensure that you can see larvae, small larvae and eggs in a solid pattern.
- Try to take a look at the queen. Once the colony is opened, the bees should be calm (especially after smoking) and numerous enough that they fill most of the spaces between the combs.
- The condition of the equipment may reflect the care the bees have received, so be suspicious of colonies in rotten, unpainted or badly painted wood.
- Ask an experienced beekeeper to check

whether there are any signs of disease.

Other sources of bees

There are a few ways of obtaining bees: beekeeping supply stores, local associations, finding a swarm, buying a nucleus, and buying a package of bees.

Many beekeeping supply stores will have hives of bees for sale, but these are likely to be new hives and therefore this

is going to be an expensive purchase.

Most local beekeeping associations have magazines in which beekeepers advertise hives, bees and equipment for sale. Beekeeping associations often have an annual auction. The equipment is mainly second-hand, and any hives with bees will have been examined for disease by an inspector before the sale, which

BUYING A COLONY

1 *A beekeeper carries a colony of bees, complete with pollen and honey.*

2 *The colony is in a working hive so it is already functioning.*

3 *The bees begin to venture out in their new location.*

4 *An entrance reducer helps prevent pests getting inside the hive during winter.*

Above: An auction is a good place to buy second-hand hives, beekeeping equipment and bees that have been vetted by an inspector, who checks them for Foul Brood disease.

Above: 'Nuc' boxes awaiting a buyer. You should get a laying queen and a small colony of bees in each box.

will help to reassure new beekeepers.

Bee swarms are gentle, but once settled in their new hive, those same bees may prove to be a strain that is prone to swarming. You will also have a queen of indeterminate age. Swarms are a risk but at least you will have a young colony (if not a young queen) that will grow as your experience grows.

Most beekeeping suppliers or specialist bee breeders will supply a nucleus (or 'nuc'), which is a type of hive for travelling that consists of four or five frames of bees with brood and a laying queen. The small colony will be gentle and easy to integrate in your hives.

A package of bees will consist of 9,000 to 20,000 bees with a queen in a small cage. Packages are purchased from bee breeders and so can be relied upon to be ready to install. However, it will take a few weeks before any brood

Above: A nucleus usually consists of a few frames of bees with brood and a queen.

BUYING SECOND-HAND

The main problem with buying your hives and bees second-hand is that you may be buying trouble in the form of disease, lack of Varroa treatment, rotten woodwork under the paint, bees with uncertain temper, a weak queen or a nasty strain of bees. If you do buy bees and hives in this way, ensure that you have a very experienced beekeeper with you to inspect everything. Annual beekeeping auctions are probably the best way to buy second-hand equipment, because at least you know that the hive occupants will have been inspected for disease.

Above: A 'nuc' box full of bees waits to be transferred to the main hive.

Left: If you get a swarm from another beekeeper, you will start off with a queen and a young colony of bees, although the queen may not be young.

TRANSFERRING BEES TO A HIVE

There are different methods of transferring and handling bees, according to how they have been acquired, but good preparation of the hive and other equipment is always essential.

If your hive is prepared to receive bees, it is possible to install a swarm, which is a ready-made colony. The hive should contain at least one box (the brood box) with one frame of comb and the other frames of foundation. Swarms are ready to make comb, so having one frame of comb isn't essential, but it will give the new colony a boost, and enable the queen to start laying at once. It also allows honey and pollen stores to be collected.

Hiving a swarm

Remove the five central frames from the box and pour the bees into the centre. Shake in any remaining bees and gently replace the central frames, close up the hive and leave the swarm to settle in. Almost immediately, you will see bees coming and going from the entrance.

A spectacular way of installing a swarm is to place a temporary ramp up to the hive entrance and place a white sheet over this. Pour the bees on to the sheet on the ramp and they will immediately start walking purposefully up the ramp and into the hive. Within a few minutes, all of the bees and the queen will have entered the hive and started to work.

The swarm will only stay in the hive if their queen is in there with them. If you see the bees gradually vacating the hive and hanging up on a branch, that's where the queen will be. You need to start again – collect the swarm and pour it into the hive. If the swarm has a marked queen, you can check if she is there when the swarm is installed in the hive.

1 *A swarm is clinging to a tree branch and is held over the open hive.*

2 *The bees will eventually disperse into the hive and start working.*

1 *A 'nuc' box full of bees is placed on an empty hive and left overnight so the bees can be transferred to the hive the next day.*

2 *The beekeeper checks to see if there is a queen present on a frame from the 'nuc' box.*

Installing a nucleus of bees

This method is common in most countries, including the UK and Europe. The nucleus ('nuc') will arrive in a box containing three to five frames of bees, including at least one frame of brood and one or two of food stores, plus a laying queen.

Many beekeepers simply place the box on top of the hive, with the entrance facing the same way as the hive entrance, and leave them there overnight. This will orient them to their new site without further disturbing them. The next day, open the hive and remove the required number of frames from the brood box. Then, gently open up the 'nuc', remove each frame of bees and place it in the hive, in the same order as they were before. With a bar of wood or an entrance reducer, close up the entrance to one bee space. This will help prevent the young colony from being robbed. You may have to feed the bees sugar syrup in a frame feeder until there is a natural nectar flow. This is one of the easiest ways of obtaining and

Above: A bar of wood is inserted at the entrance of the hive to reduce the space to one bee space. This helps a young colony to guard against robbing.

Right: A bee exposes the Nasonov gland and emits a pheromone that signals to other bees to move into the new hive.

installing bees. You have a young colony with existing brood, headed by a young laying queen, that will grow as you gain experience. It is easy to buy, easy to install and will give you confidence in managing your bees, because small colonies are usually mild-tempered.

Above: If the hive is baited, a swarm will enter of its own accord.

BEE SIGNPOSTS

However you hive a swarm, there will be many bees swirling around in the air that are not quite sure which way to go. Forager bees will immediately take up position at the hive entrance, and, facing toward the entrance, they will rapidly fan their wings, exposing a small gland (which shows white) down near the sting. This is called the Nasonov gland, and it emits a pheromone which says to all the circling bees: 'Here we are. Here's our new home, come on in.'

Above: It is useful to have a queen marked (with a dot of paint) when you buy a nucleus, so that you can locate her easily.

Installing a package of bees

This is popular in the USA, and is slightly more complicated than installing a 'nuc', but it is still a relatively problem-free method of obtaining bees. Prepare to install your bees in the late afternoon. This will inhibit robbers such as wasps, since the day will probably be cooler and your bees will be more inclined to settle down and not drift away. Make sure that the package or packages of bees are kept in a cool, shaded place until you are ready, then carry out the following procedure.

Set up a floor with one hive body (the brood box), and remove five of its frames. Make some sugar syrup (one part sugar to one part water), put it in a small, clean garden sprayer and spray the bees through the screen; the bees

Above: Spraying the package of bees with a sugar syrup ensures that they are replete and sticky, so that it is a simple task to pour them into the hive.

will gorge themselves with the syrup and become sticky, so that it is easy to pour them out into the hive in a clump.

Remove the package lid, remove the can of syrup provided for transit, find and remove the queen, suspended in

Right: Use smoke to check that the new queen is in good condition.

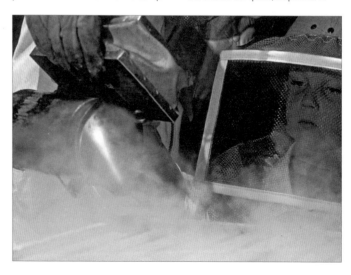

Above: The bees will get used to the queen in the cage and begin to feed her.

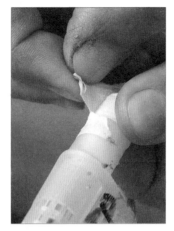

Above: If the queen is fine, remove the cage and take away the tape.

Above: The queen has escaped from the cage. If she is marked, you may find her.

her cage, and re-close the package. Replace the lid. The queen cage has holes at both ends plugged with cork, and one end is visibly filled with white 'queen candy'. Remove the cork from this end and suspend the queen cage between two centre frames in your hive. Workers will eat through the candy and gradually release the queen.

Gently and lightly shake the package so that the bees fall in a group to the bottom. Remove the lid and shake the bees into the hive on top of the queen. As the bees move through the hive, gently return the frames you removed earlier. Close up and feed your bees as required, using the same feeder or a frame feeder until the nectar flow begins.

After a couple of days, check that the bees have released the queen from her cage. If they have not, you can now safely release her and allow her to move on to the frames. One week later, check the colony once again to ensure that the queen has started laying eggs, and that her brood is developing.

Right: Until the nectar begins flowing, the bees can be fed using a vacuum feeder.

BEEKEEPING CALENDAR

Beekeeping is very much a seasonal affair and the bees operate to a rhythm dictated by the natural world, which will change only according to the bees' priorities; beekeepers should simply attempt to ensure that the bees are well housed and kept disease-free as far as possible. The beekeeping calendar is centred around nectar flows in the local area, and you need to find out about the nectar and pollen sources; it is an essential part of beekeeping. The next few pages will guide you through a beekeeping year and help you make the most of your colonies.

Left: This colourful garden is filled with a variety of plants that provide a rich source of nectar and pollen for foraging bees.

SPRING

This is a time of rapid build-up for honey bee colonies, when much of their honey will be collected and stored, and reflects the fast spring growth of nectiferous flowering plants.

Spring is the season when the beekeeper should inspect the growing colony or colonies regularly, keeping a wary eye out for the problems that can occur at this busy time, especially swarming. It is best to take an experienced beekeeper with you for your initial inspections.

Inspecting the bees

Before you inspect any hive, light the smoker and have your hive tool ready. Carry spare smoker fuel and matches or a lighter. As you approach the hive, stand to the side out of the bees' flight path, and gently puff some smoke into the entrance. Then carefully lever off the lid and puff some smoke over the tops of the frames to keep the bees' heads down. If there is a honey box, remove it and the queen excluder so that you can inspect the brood box.

What to look for in the brood box

Look for the queen and check that she is laying eggs. Look carefully for tiny stick-like items in the base of the cells.

Above: The temperature inside the hive should be 32–36°C/90–97°F.

Ensure that there is brood of all stages present, and make sure that there is no sign of disease and that the hive floor is clean and not covered in debris. Check that the colony has sufficient food – both nectar and pollen.

Try to determine if the colony has built up since your last visit, and make sure that there will be sufficient room for further expansion in the period before your next visit.

How to inspect the frames

First, remove the frames at the edge of the box to give enough room to separate the other frames. Place one of these frames on to the lid on the ground. Then gently force the other frames apart and take out the centre frame. Study this for capped brood, larvae, eggs and stores. You may even see the queen on one of the brood frames, in which case, gently lower the frame back into position. Keep lifting out and studying the adjacent frames until you have a good idea of the extent of the brood and stores, and remember that if you have seen eggs and brood, it is highly likely that you have a good queen somewhere in the hive. If you notice anything strange such as queen cells or a bad smell, ask the beekeeper

Above: Spring is a good time to clean the hive floor with a blowtorch.

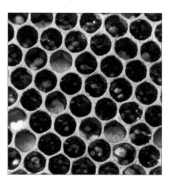

Above: Check that there is sufficient honey to sustain the colony.

Above: If you see queen cells, it may indicate that the bees will swarm.

accompanying you for an explanation. Make sure you have checked for disease, or signs of swarming. Now is the time to carry out any *Varroa* treatment required – a vital precaution. We take a look at the various treatment options available in a later chapter.

Swarming

The rapid build-up of the colony in spring may soon lead to swarming, in which half or more of your bees can depart the hive under a new, young queen. With more bees in the hive, the queen substance, pheromone, becomes less concentrated in the hive and workers may start the construction of queen cups along the bottom edges of brood frames, which can develop into queen cells. This is often the first indication to a beekeeper that swarming is likely to occur soon, or has already occurred. Destroying these queen cells as a method of swarm prevention doesn't usually work – you are likely to miss one, or the bees will have swarmed anyway. Always try to prevent swarming before the impulse starts.

Above: Swarming may occur if the hive becomes overcrowded.

Above: Look at the frames of stores; if there is no honey, feed the bees.

Above: A honey bee with laden pollen baskets lands on a Shasta daisy.

Above: Listen carefully close to the hive. If all is well, you should hear a healthy buzzing. If it is quiet, there may be a problem inside the colony.

SPRING SWARM CONTROL

Remember that a swarm departing from the hive will take a full month out of the colony's productive period, so try and prevent this happening by working with your bees.

There are several manipulations that may limit or prevent swarming, but none are guaranteed to work and some may well interrupt the collection of nectar. You can try the following:

• Replace the queen at least every two years. New queens swarm far less often.

Above: When the bees swarm, they divide into two colonies.

• Use two brood boxes to give more room in the nest area and more space for the queen to lay eggs.
• Bees work upward and tend to crowd

the upper brood box if you are using two. Simply keep reversing them as required.

• Give the bees adequate honey storage room by adding new supers.
• If one of your hives becomes very strong, relieve any overcrowding by moving some of the frames to a weak hive, first making sure that the weak hive isn't suffering from disease.
• Ventilate your hives by keeping the entrance fully open. You can also paint the tin lids white to reduce heat in the hive. Good ventilation goes a long way in reducing the swarming impulse if other methods are also employed.

Making an artifical swarm

If a hive does become crowded and queen cells are formed, then the colony may have already swarmed. If it hasn't, you can easily divide it and make an

Above: Adding another brood box gives the queen more room to lay eggs and produce young, and reduces overcrowding.

Above: Adding more supers allows the bees more space to store honey and also helps reduce overcrowding.

Above: If the hive is overcrowded, another option is to move some frames to a weaker hive.

COLLECTING A SWARM

1 *This swarm has collected on the branch of a tree so it is relatively easy to catch.*

2 *The beekeeper cuts the branch and has a bee basket ready to take the swarm.*

3 *The swarm is gently lowered into the basket, then taken to a hive.*

artificial swarm to make the bees think they have swarmed. Move the hive to a new position anywhere in the apiary. Place a new brood box on its floor in the old position, and put the queen on a frame of brood in this new box. Fill this new box with foundation or comb.

Place the original supers on the new hive and cut out the queen cells in the old hive. A week later, cut out any others but keep one or two. The bees will decide which cell to use, or, get rid of all the cells and introduce a new queen in a cage, or add a queen cell from another hive.

These methods will reduce or eliminate the swarming impulse in the hive, but you will also increase your stock. If you decide later that you don't want this extra colony, then unite it with the original colony and let the two queens fight it out, or destroy the queen you don't want and unite the colonies.

Finding the queen

If you can't find the queen, cut out all of the queen cells. Split the colony in two, ensuring that each half has both eggs and brood frames, and give each

half a floor and a lid. In three days, the half with eggs will have the queen. The other half will not have a queen, but may by now have queen cells. In the queenless colony, cut out all of the queen cells except one or two and let them produce a queen. If you don't want a second hive, cut out all of the queen cells and unite the two halves again under their original queen.

This method of destroying the queen cells will only be successful if you are certain that you have removed 100 percent of them.

Above: Remove any queen cells that you see, using a hive tool.

Above: A beekeeper checks supers filled with honey.

Above: Divide the hive to prevent swarming; introduce a new brood box and frames.

SUMMER

Once the frantic activity of spring is over and the chances of swarming are reduced, it is time to ensure that your bees increase in numbers to take advantage of any major honey flow.

You may already have experienced an early honey flow, depending on your location, and be in a position to harvest some of the honey. However, the main task now is to ensure that nothing goes wrong with the colony or the hive. You should ascertain that there is sufficient storage room for the honey, and keep checking for disease and other problems.

Before you decide to remove any spring honey, make sure that the bees have sufficient stores to last them over the late spring and early summer period, when there are few, if any, flowering plants available. You could end up taking their stores and leaving them to starve just before a major flow of honey.

Keep a careful eye on your hives during the summer period and maintain a regular inspection schedule as you did in the spring, using the same checks.

Inspecting a closed hive

Checking the hives from the outside is not a substitute for regular open hive inspections, but each time you go past your hives, simply look and see what is happening around them.

When observing the outside of the hive, look out for the following:
- Fighting at the hive entrance, which suggests robbing.
- Bee larvae on the alighting board, which suggests that there is no food.

Above: After harvesting the honey, you may need to add more supers if there is another honey flow.

Above: A beekeeper needs to wear gloves and a veil when inspecting the hives to check that all is well, for example, that the bees are producing honey, and to confirm that there is a queen and brood present.

Above: Bees that appear mummified have suffered from chalkbrood disease.

Above: Many dead bees suggest that they have been poisoned or they are starving.

Above: Once you have harvested the honey, you can reuse the frame and comb.

- Drones being evicted from the hive by workers.
- White, hard, mummified larvae at the hive entrance, suggesting an outbreak of chalkbrood disease.
- Heavy faeces spotting on the hive, especially around the entrance, suggesting bee dysentery.
- A pile of dead bees at the entrance, indicating either pesticide poisoning or starvation.
- Ant columns or many wasps entering the hive, suggesting a failing or dead colony.
- Bees issuing in their thousands from the hive and swirling up into the air, suggesting a swarm. However, many

bees flying around and facing the front of the hive simply means that young adult bees are orientating themselves on their first flight.
- If you see busy flight activity around all of the hives except one, inspect it immediately.

Drawn-out comb

If you provide drawn-out comb (wax that the bees have already formed into cells) in the honey supers rather than new foundation, the bees will accept this and immediately begin to store nectar in the drawn-out comb, thus saving them a lot of time and energy; they will also not need to use up the honey stores to give them energy. Drawing out wax comb from foundation requires a lot of honey: it takes 6–8 kg (13–18lb) honey to make 1kg (2.2lb) wax.

Usually, you will only have spare comb in the supers during your second season, but after you have taken the honey harvest, you will be left with several frames from the supers with drawn-out comb that you can reuse for the next season.

Above: If wasps gain entry to the hive, it may indicate a failing colony.

MARKING THE QUEEN

Beekeepers often mark their queens with a blob of paint on her thorax, making them far easier to find. A commonly used colour/year code tells you which year she was born. Devices to assist in marking your queens can be purchased at most bee supply stores.

Year ending	Colour
0/5	Blue
1/6	White
2/7	Yellow
3/8	Red
4/9	Green

MOVING BEES IN SUMMER

Many beekeepers will be content to keep their hives in one area for the whole year, but it sometimes becomes necessary to move them during the summer period.

Above: A pick-up-truck is ideal to move beehives. The hives have ventilation through the roofs, with screened entrances and roofs.

Above: Flatbed lorries are ideal for moving large numbers of hives easily.

Above: Any movement of bees or honey stores is often done just before twilight.

Moving bees is hard work and can be disruptive for the bees. However, in rural areas, if a crop fails, or if there is to be insecticide spraying anywhere nearby, beehives may need to be moved to a new site to avoid being affected. A knowledge of how to move them so that the colony continues to thrive is vital, and will also help with swarm collecting in future years.

If you are moving a hive just a short distance of less than 1m (3ft) in one direction this will initially confuse the forager bees, but they will generally be able to find their hive again.

If you intend to move the hive a greater distance than 1m (3ft) but under 3km (2 miles), problems will occur because the forager bees will know the area they are in and will navigate directly back to their original hive site, where they will gather in a small cluster on the ground, unable to find the hive.

Above: Moving bees can give the beekeeper access to more nectar sources and greater honey yields. This beekeeper in Greece has harvested honeydew from pine and fir trees.

Above: In Greece, hives are moved to forests yearly to increase honey production.

Above: In the UK, hives are often moved to fields of beans and oilseed rape.

Above: Many beekeepers take their bees to the heather moors. These bees are being transported in a horse box, which has a tailgate that can be dropped down; the hives can then be moved on a barrow.

Above: Amateur beekeepers may use the family car to move bees; this can be dangerous. However, the hives are safe if they are enclosed in special tough, ventilated bags.

COOLING THE HIVE

During the summer period, on a hot day, you may see the bees fanning at the entrance to the hive. Honey bees have extensive thermo-regulatory abilities, and in order to keep the brood nest at 34°C (93°F), they will fan the hive inside and out to cool it down. In very hot weather, they will bring water droplets into the hive and fan these to lower the temperature further. By painting the hive lid white in hot climates, you will help keep it cool in hot periods, and the bees will spend their time collecting honey, rather than water.

Above: If you are moving your bees on a hot day, spray the hives periodically with water to keep them cool.

Moving a long distance, for example to a new site that is outside their original forage range and more than 3km (2 miles) from the original site, the bees will not recognize the new area and its landmarks, and will orient themselves to the new site without difficulty.

Moving strategies

There are ways of overcoming moving problems: for example, by moving a hive 0.6m (2ft) each day until the required distance is reached, or by collecting bees returning to their old site in spare boxes. Overall these solutions are not very effective in time management. When moving hives a short distance, some beekeepers will block up the entrance with grass. By the time the bees have chewed their way out over a day or two, they will then orient themselves to the new site. This can work – but you will still lose some bees.

When to move bees

Bees can be moved in any weather, and at any time of day or night. However, there are certain points to bear in mind:

• If you move bees during the day, keep them confined over the previous night and do not open up the hive until you have completed the move.

• If you confine bees for more than two hours, you should always provide them with more space in the form of an extra empty box, and a good degree of ventilation by using a gauze lid. If it is a hot day, provide them with plenty of water by spraying it at regular intervals through the gauze.

• If the weather is very cold or it is raining heavily, you can just load up and go. The bees will stay in or on their hive. The same applies at night.

AUTUMN

While there is a honey flow, the bees are usually good-tempered, but by the end of summer you will have harvested the surplus honey and the bees may become more aggressive.

There are many useful tasks that you can carry out in autumn, such as reducing the hive entrances to help the bees defend their nest. Wasps are a particular nuisance at this time of the year, and it is well worth destroying any wasp nests in the vicinity.

Remove the queen excluders; these are no longer needed over the winter and if you do have a honey box on the hive, an excluder will hinder the bees from accessing the stores.

Take this opportunity to remove any odd-shaped or defective combs, or broken frames. If they still contain stores, place them to the side of the box until they are ready for removal. Old black combs should be removed for rendering down and replaced in the spring with foundation or with new comb.

It is important to ensure that the lids and the floors are sound. Broken or leaking lids and floors can be the direct cause of colony demise over the winter period. Ensure that the hive stands are up to the job, as they may need to withstand winter floods.

ESTIMATING HONEY STORES

- A Langstroth frame holds 3kg (6.5lb) of honey
- A shallow Langstroth frame holds 1.9kg (4lb) of honey
- A British Standard frame holds 2.5kg (5.5lb) of honey
- A Shallow British Standard frame holds 1.6kg (3.5lb) of honey
- A Modified Dadant frame holds 3.9kg (8.5lb) of honey

Above: Puff a little smoke into the hive and remove the queen excluder, if used.

Autumn inspections

These should cover the same points as the spring and summer inspections, with the exception of looking out for signs of swarming.

If you find a weak stock, first make sure it is not diseased, and then consider uniting it with a stronger stock. You may have a queen that is not laying well, or a bad strain of bees. Kill the weak queen first by squashing her, otherwise the two queens will fight, and inevitably the weak and useless queen will win! Try not to be squeamish about killing weak or ineffective queens. It is for the good of the bees that you do so.

Above: Record the condition of the queen and when she should be replaced.

Record the age and condition of the queen. She should still be laying, although this will be at a reduced rate compared to her performance in spring. If she is two years old, she should be marked for replacement either now or in the spring.

There are advantages to re-queening in both periods, which we will examine later in the book.

Treating for Varroa

You should treat the colony for Varroa in early autumn. It is worth alternating the treatments, for example, using a chemical treatment in the spring and an organic treatment in the autumn. This will limit the development of resistance.

Autumn feeding

Late autumn is the time to feed your bees for the winter period, especially in cold climates. Use sugar syrup in a frame feeder to feed your bees .

As a general rule, the bees need the following stores to help them over the winter period:

Cold northern climate A minimum of 40kg (88lb) in three boxes.

Temperate climate 15–30kg (33–66lb) where the average winter temperature range is -4°C (39°F) to + 10°C (50°F). This includes the UK, New Zealand and western Europe.

Above: A frame feeder filled with sugar syrup is inserted into the hive.

Warm southern climate 8–15kg (17.5–33lb) where the average winter temperature is above 10°C (50°F). Each 5 litres (1 gallon) of sugar syrup will increase stores by 3kg (6.5lb). For autumn feeding, use thick syrup: 1kg (2.2lb) sugar to 500ml (1 pint) water.

Above: Use a bee fork to look for Varroa mites on a frame.

Above: Autumn is the second time of the year to add anti-Varroa treatment.

Above: Check the floor of the hive for the Varroa mite.

WINTER

Now that you have made sure your bees have all the stores they will need for the cold weather that is coming, you can settle them down for winter.

As long as you have carried out the necessary autumn tasks, and ensured that the bees are well housed and have sufficient stores, you should now leave them alone unless you need to respond to a crisis. You should still visit the hives regularly to ensure that all is well.

Winter inspection

Observe all the hives to check if the lids have blown off. Use bricks or straps to keep them on. Sometimes hives blow over, so in windy areas, use ropes to tie them down. Inspect the hives for flooding and ensure that the entrances are not blocked by snow.

Check that the entrances are sufficiently reduced to prevent mice from entering. Use mouse guards or

Above: use a brick to hold the lid firmly on top of the hive.

Above: Check regularly to ensure the entrances are not blocked with snow.

Above: A mouse excluder has small holes to allow only bees to enter the hive.

Above: Polyurethane hives have good insulation and will keep the bees warm and dry, so that the bees have a good chance of surviving the winter.

Above: Fit a mouse guard to prevent small creatures entering the hive.

mouse excluders to protect the entrance, and make sure that they are securely fixed to the hive. Go over each hive to ascertain that all cracks and holes in the hive are sealed, but there should be some ventilation and hives should be tilted slightly to allow water to run off.

In some areas, especially subtropical areas of Europe and the USA, the queen may well continue to lay eggs right through the winter period, and the bees may continue to gather nectar if there are suitable plants available. In colder areas, there will be a quiet time when the queen ceases egg laying, which lasts until soon after Christmas, or later during a very severe winter.

Above: A mouse excluder is shown fixed to a Dadant beehive.

Warming the hive

The bees' legendary powers of thermo-regulation again come into play over the winter period. In order to keep the brood nest at 34°C (93°F), they will, if necessary, form a bee cluster inside the hive. The bees begin to cluster at around 18°C (64°F), and the formation will attract more bees and become tighter at around 13–14°C (55–57°F).

The cluster surrounds the brood and maintains it at the correct temperature. There are always passages for the movement of bees in the cluster, and it will expand or contract depending on the temperature of the hive. By shivering, the bees generate heat within the cluster. The bees on the outer part of the cluster are colder than those in the centre and may become semi-dormant, but if there is a threat they will still be able to extend their stings.

Bees can only maintain warming if they have access to stores, although in some very cold periods the bees can die even if there were stores in the hive. It is damp, however, rather than cold, that is the major killer of bees in the winter.

Some beekeepers fear that a mesh hive floor will make the hive too cold for the bees during the winter, but the bees keep the brood nest warm and do not attempt to heat the whole hive. After all, in nature bees may nest in a tree hollow, with no floor below. It may be healthier to use a mesh floor because debris such as dead bees or dislodged *Varroa* mites will simply fall through.

Preparing for spring

During the cold winter months, there is time to ensure that you have all the equipment required for the return of spring, such as supers of comb, frames of foundation, spare boxes, clothing and tools, sugar syrup for early feeding and pollen patties for protein feeds.

Above: A beekeeper places an Ashforth feeder above the brood nest and below the roof. The feeder can be refilled with sugar syrup without disturbing the bees.

FROM WINTER TO SPRING

In mid- to late December, in climates that experience changes in seasons, the queen will again start laying eggs in the warm confines of the brood nest and the colony returns to life.

Once the bees start waking up after the winter and the queen is laying eggs, it signals the start of another hectic, expansionary spring, and you will need to start a regular routine of inspecting the hives again. As the daytime temperature gradually increases, the warming cluster of bees will begin to break up. It very much depends upon the climate, the weather and the geography of your area, but a colony can suddenly explode in number in early spring and take the beekeeper completely by surprise, so be prepared.

First inspection

On a fine day, in winter in the northern hemisphere, when the bees are flying strongly, make your first inspection of the hive. Check for the following:
Look for evidence that the queen is alive and laying. If the queen is not laying, or is dead, unite the stock with another healthy stock. Ensure that there is brood of all ages, and that there are stores of both honey and pollen. Examine the hive for any signs of disease. Always look for disease in all inspections.

Above: In colder climates, snow is common and the bees may need a little help with feeding until they can get out of the hive to forage for their own supplies again.

Feeding with sugar syrup

If the bees need stores, feed them now. For spring stimulation or pollination feeding, use a thinner mix of 1kg (2.2lb) sugar to 1 litre (2 pints) water. Small feeder buckets can be used, or even large jars. Simply puncture the lids with small holes, fill with syrup and turn

Above: Ensure that there are stores of pollen in the hive.

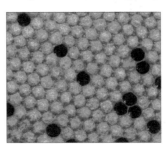

Above: The presence of brood is a good sign after the winter period is over.

Above: Rock roses are an early flowering plant, and a good source of pollen.

upside down over the top frames. Place an extra empty box on the hive to cover the feeder, and put the lid on top.

The use of invert sugar syrup in the spring feed gives the bees a surge of energy. When they collect nectar or take sugar syrup, the bees invert the sugar: they turn the disaccharide sucrose into its two component sugars, glucose and fructose, with the addition of enzymes that enable the bees to better pack the sugar in the cells. By feeding them invert sugar, you will save the bees energy and time, and give them a good spring boost.

Feeding with pollen

Bees need pollen as their protein food, and colonies that have insufficient supplies will dwindle and die out, or at best grow very slowly. If there are no early pollen sources in your area, you should feed pollen to the bees in the early spring. Pollen patties can be made or purchased. The best method of feeding is to spread a patty in a paper bag over the top of the frames. The bees will rip the paper up to reach the patties. The requirement for pollen is often underestimated or even forgotten about by beekeepers, who then ascribe the failure of a colony to disease or queen failure.

Early sources of pollen are gorse, willow and rock rose, among other plants. Make sure you learn about the many nectiferous and pollen-producing plants that grow in your area..

POLLEN SUBSTITUTE PATTIES

You can make your own pollen substitute patties by mixing 1 part sodium caseinate with 2 parts dried non-active yeast, and adding sugar syrup to make a stiff paste.

Above: A handmade bee feeder can be made from some frames filled with sugar syrup, placed in a shallow box and put on top of the hive.

Above: A feeder bucket is filled with sugar syrup, the lid is punctured with small holes, then it is inverted on top of the hive, causing a vacuum. The bees can reach the food.

YEAR PLANNER

In any kind of farming enterprise, the year planner is often altered quite considerably by the weather, and beekeeping is no exception – simply adjust your routine and use common sense.

EARLY SPRING
- Watch out for overcrowding and queen cells.
- Observe to see if pollen is being carried in by the foragers.
- Start queen-rearing plans later in the month if the weather proves to be suitable.
- Prepare and place bait in the hives to catch a swarm if needed.
- Every time you pass your hives, carry out a quick external inspection. If something looks wrong, check it out.

MID-SPRING
- This is often the beginning of boom time in the hives, so be extra vigilant at this time.
- Continue queen-rearing activities.
- Keep a close eye out for swarm preparations.
- Commence swarm prevention manipulations, such as reversing hive bodies and adding an extra brood box if you have overwintered with only one.

LATE SPRING
- Increase inspections to once every seven to ten days.
- Replace your queen excluder if you add honey supers.
- Treat for *Varroa*.
- Start re-queening your hives if necessary.
- Add supers so that the bees have enough storage.
- Keep up with swarm prevention manipulations and watch for swarming preparations. Carry out an artificial swarm.

EARLY SUMMER
- Keep adding more honey supers as necessary.
- Extract spring honey if required; however, unless you know that there will be sufficient sources of honey to keep the bees supplied adequately, you should leave this honey in the hive for the bees' own use.

MIDSUMMER
- Check to see that there are enough honey supers for the increased honey flow.
- Continue to look for signs of swarming and maintain all the swarm prevention techiques that have been mentioned earlier.
- If necessary, carry out an artificial swarm.
- Order new queens now if you want to re-queen with purchased queens in the late autumn.

LATE SUMMER
- Remove the main honey harvest.
- Start queen-rearing activities if you want to re-queen in the autumn.
- If you need to move your hives to winter sites, do it now while the weather is clement.
- Treat for Varroa mites.
- Clean the empty honey boxes and carefully store them away after treating the comb with a preparation against wax moth.

Above: Papaver rhoeas

Above: Hyacinthoides hispanica

Above: Taraxacum officinale

Above: Aquilegia

Above: Berberis

Above: Eryngium giganteum

EARLY AUTUMN

- Thoroughly inspect the hives in preparation for winter. Check all the woodwork, lids, floors and frames to ensure they are sound and there are no holes anywhere.
- Reduce to two brood boxes only, as this will not be a productive period for the bees.
- This is the best time to re-queen if that is required. (Most new or hobby beekeepers will re-queen in the spring.)

MID-AUTUMN

- Prepare your hives for winter. Feed the bees with sugar syrup if necessary, and ensure they have sufficient stores to last them for a few months.
- In windy areas, tie the hives down or put a brick on the lid.
- Place mouse guards over the entrance at one bee size, so that larger animals cannot enter.
- Ensure that the hives are on ground that cannot be flooded.
- Tilt the hive slightly so that water will run out and off.

LATE AUTUMN

- Take a look at your hives whenever you can, especially after bad weather conditions such as wind, rain or snow.
- Check for animal damage such as wax moth caterpillar, or invasion by pests such as mice, raccoons or even larger creatures such as bears.

EARLY WINTER

- Clean old comb off frames and render it down.
- Prepare new frames of foundation for spring.
- Repair and clean any old equipment such as the hive floor and lid.

MIDWINTER

- Ensure you have sufficient boxes and frames of comb or plastic foundation to cover the coming year.
- Clean up equipment and repair any damaged items, including unused floors, lids and boxes.
- Check all your stored comb for signs of wax moth.
- Pests will become more active as the weather begins to warm up.
- If you want to purchase queens, ensure you place your orders early. If you prefer to produce your own queens in the spring, collect all the necessary equipment now.

LATE WINTER

- If the weather permits, carry out your first hive inspection of the year.
- Ensure that the bees are in good condition, and that they have sufficient stores of pollen and honey. Feed them with sugar syrup if necessary.
- Remove mouse guards; replace with entrance reducers.

Above: Echinops ritro

Above: Ilex aquifolium

Above: Fuchsia

HARVESTING HONEY

For the beekeeper, the honey harvest is the highlight of the beekeeping year. It is also the culmination of a lot of work involving the application of hard-learned beekeeping skills and knowledge. Bees can make an astonishing amount of honey; yet each bee collects an amount of nectar only the size of a pinhead before returning to the hive. Collecting the honey from the hive can cause new beekeepers some anxiety, but with the correct equipment and a plan of action, the whole process is quite easy, if a little sticky! It is also great fun, and a good way to involve the whole family.

Left: The final product of rich golden honey on the comb is worth all the hard work of the bees and the beekeeper.

EXTRACTION EQUIPMENT

Before you consider harvesting your honey, make sure that you have all of your equipment ready and immediately available, and that you know the correct steps to follow.

It is pointless taking boxes of combs dripping with honey out of the hive unless you are able to extract it at once and quickly clean up, so make sure you are properly prepared. There are several reasons for this. First, the combs will make a mess – honey will drip from them and land on the floor, where it will attract ants and other small predators, or you may even slip on it. Second, bees will be after that stolen honey, whether they are your own bees or neighbouring bees, and will ceaselessly attempt entry to your house or shed – and remember, if one gets in and makes it back to the hive, thousands more will appear! Third, harvesting honey takes up space, usually in the kitchen, which is probably also needed for family use, so the quicker and more efficient the process, the better.

Major requirements

A bee-proof room is essential. It should preferably have running hot and cold water and plenty of sink space. Unless you have invested in a honey-extracting plant of some sort or kitted out a shed with the necessary equipment, this is usually the kitchen.

You cannot harvest honey without an extractor, so this piece of equipment is a necessity. You can buy one new or second-hand, but ensure that it is made of food-grade plastic or stainless steel. Older extractors may be made of tin plate and shouldn't be used.

Many beekeeping associations have a shared extractor which is passed around the members, or you could

Above: Some clear honeys and honeycomb. The colour of the honey is dependent on the plant source, and can vary from very pale to very dark.

assist another beekeeper who has an extractor, in exchange for the use of his or her machine. In some areas there are even professional beekeeping operations that will extract your honey for you for a price, but that spoils the fun for the hobby beekeeper, who usually prefers to do the job himself.

Extractor design

Designed to spin out the honey either tangentially or radially, for a small-scale operation either type of extractor will do. Most hobby extractors usually hold about four frames at a time and can be hand-driven or motor-driven. A four-frame extractor is ideal for the

Above: The wax cappings are removed before the honey is extracted. Keep the wax for use in the future, to sell or make into candles.

Above: A radial extractor spins the frame and the honey is flung out from both sides of the frame at once.

Above: With the tangential extractor, after one side of the frame has been extracted, the frames need to be turned around.

Above: To use this hand extractor, the frames are simply rotated by hand, using the handle on the outside.

beekeeper who has just a few hives. Extractors with motors are much easier to use and are not that much more expensive. They may also extract more honey, but many hobby beekeepers are

quite happy with their hand-driven models, at least to begin with.

The extractor needs to have a wide base platform to help stabilize it when it is spinning. This is more important with

the motor-driven ones because they operate at higher speeds, and extractors carrying even slightly uneven loads can dance all round the room unless they are well stabilized.

OTHER HARVESTING EQUIPMENT

Once you have established an extraction room and obtained an extractor, you must now collect together some more items in order to harvest the honey efficiently.

The following items are vital parts of the honey harvesting process. These can be as cheap or as expensive as you wish.

Uncapping knife or fork

This implement can simply be a long bread knife. The ones with a serrated blade are ideal. Some beekeepers use an uncapping fork, which can also work well but is best used for very uneven comb, when the knife may miss cells of honey. You can purchase electrically heated knives or steam-heated knives,

which are very effective. If you want to splash out, there are other expensive mechanical devices, such as brush uncappers, which gently brush off the wax cappings, leaving the comb intact.

Large bowl

This is to hold very hot water to heat the uncapping knife (or bread knife), unless you are using a heated knife or other device. It is much easier to uncap comb cells using a hot knife. Top the bowl up with hot water as you work.

Large container

This is to store the honey after you have emptied it from the extractor, and must be made from either food-grade plastic or stainless steel. An extractor can only hold a limited amount of honey before it requires emptying. Storage buckets containing 20–25kg (45–55lb) are ideal. It is even better if they have a tap near the bottom so that you can easily decant the honey into jars or other smaller containers later on. Most honey will set in time, and it is

Above: This beekeeper is about to use a heated uncapping knife to remove the wax capping from the frames containing the honey.

Above: Strong-flavoured honey tastes better if it is blended with a lighter variety.

Above: There are many designs of uncapping knives available.

Above: Some people prefer a heated electric uncapping knife for speed.

Above: Heating the knife in hot water makes uncapping a lot easier.

far easier to decant the honey soon after extraction, rather than having to spoon it out later when it has set firm. Always ensure that honey buckets have their taps firmly shut before you start; the honey flow is silent.

Filter device

This can be almost anything that will filter out the large bits and pieces that will be in the honey: bits of wax, dead bees, the odd wax moth, bits of broken frame and so on. The filter should be placed between the extractor and the storage bucket. How fine the filter is depends on how much debris needs to be removed. You can purchase filters of varying sizes and types from bee supply shops, and these are simple to use.

Some commercial companies use high-pressure filters containing diatomaceous earth, which removes just about everything from the honey. Some beekeepers use a sieve with a muslin liner to provide a finer filter than a sieve alone. The problem with this is that the finer the filter, the slower the operation becomes.

Floor trays or newspaper

Your boxes of frames will drip honey as they sit, waiting to be extracted. To avoid getting honey all over the floor, it is best to place the boxes on trays or paper both before and after extraction.
Large container This is needed to hold the cappings that have been trimmed from the frames of honey. The container

will have to be emptied regularly into a large spare bucket, before straining off any honey from the cappings.

Above: Until you are ready to bottle your honey, a well-sealed container is useful.

Above: The tank valve on this bucket makes it easy to decant the honey.

Above: This simple filter comes in different types and mesh sizes.

HONEY HARVESTING FIELD EQUIPMENT

Bee escape boards, fume boards and bee brushes are devices to help remove the bees from the supers containing honey, before you take them from the hive.

Before you begin to harvest the honey, it is prudent to collect all the equipment you will need before you begin. That way, the bees will not get so agitated when you take their stores of honey.

Escape board

Sometimes known as a clearer board, an escape board is simply a board that sits underneath the box(es) of honey combs you wish to remove. The board contains a one-way bee valve set in it. Escape boards control movement of bees inside the beehive and to the outside world. They allow the bees to move from one area to another but prevent them returning in the opposite direction.

Above: A white cottage beehive looks picturesque in a garden, along with some purple lavender. Although attractive, this type of hive is very heavy once full of honey.

Above: A one-way valve allows bees to leave the honey super, but not return.

The way they work is that at night, bees will tend to move down through the valve to the brood nest and will not be able to go up again, thus ridding the

Above: This disc can help ventilate the hive or close to prevent the queen leaving.

honey box of its bees. In warm weather, some bees might remain in the honey supers. These may have to be removed with a brush as you start to harvest the honey. Place the escape board on the hive at least one day before you take off the honey, to give the bees enough time to exit the honey supers overnight. Lift off the boxes from which you want to extract honey, and put your escape board in place over the remaining boxes. Then re-assemble the hive. If any bees are still remaining in the honey supers, use a special bee brush to remove them from each frame on the day you are harvesting the honey.

Bee brushes

These are soft brushes that are used to brush bees off the comb, without damaging them, into the box below. They work well but can become very sticky with honey, and can at times irritate the bees. Carefully use a bee brush to remove all the bees from the honey frames or to remove any remaining bees if you have used an escape or fume board.

A wheelbarrow or hand barrow

Full honey boxes are heavy, and if you need to get the boxes back to the house or shed and you are not using a vehicle, these items can be life-savers. Choose a sturdy barrow that is weather-proof and that will withstand rough treatment.

Fume boards

These are similar to hive lids, but the underside of the fume board contains an absorbent cloth (usually but not always black) that can be impregnated with bee repellent liquid. When the cloth is placed over an open box of comb, the repellent sends the bees down rapidly to the lower chamber.

To use a fume board, remove the existing hive lid, add the repellent liquid to the fabric of the escape board and place it on the top of hive for a few minutes. These boards are

Above: A wheelbarrow is a useful appliance to transport heavy honey supers from the garden or field into the extracting room.

placed on the hive on the day you wish to remove the honey supers for harvesting. Once the bees have moved down, you can remove the honey super and place the board on the next box down, at the same time clearing the super for removal and extraction of the honey. Use the bee repellent as instructed on the container; too much will confuse the bees. Instead of moving away, they will cling to the comb and they will be impossible to shift.

Above: A beekeeper gently wields a bee brush to remove any stray bees from a frame before extracting the honey.

HARVESTING PLAN

The honey harvest can be carried out at the end of each honey flow, but you have to be sure that you know in advance exactly when that will be.

Many factors determine when to harvest the honey; the weather is important, as are nearby trees, plants and crops.

Assessing the honey flow

It is not difficult to spot when your bees are working a good flow. The ceaseless bee activity and the rapid arrivals and departures from the hive are good indicators of a flow, and it is important that you take a look at the honey boxes to see what is happening. If you see busy activity, always take a look in the hive.

Inspecting for the harvest

You will be able to assess whether to harvest or not by inspecting the frames of comb in the honey supers. Follow this simple inspection procedure:

- Gently lift the lid off the top of the hive and smoke the bees.
- Prise out some of the honeycomb frames and inspect them.
- Note how many frames of capped honey are present.
- Go through the other boxes above the brood boxes in the same way.
- Check each frame and ensure that at least three-quarters of the cells are capped with wax. You can only extract honey from these frames. Honey from frames with less than three-quarters of the cells capped will contain too much water and may ferment.

Whether to extract?

Having carried out your inspection, you need to decide if it is worth extracting the honey. If there is only one box of

Above: A frame is lifted out of a honey super so that the beekeeper can inspect whether the cells are capped.

Above: The beekeeper gently shakes a frame to remove the majority of the bees to assess the amount of comb present.

Above: Each frame is checked to see if there is capped honey.

Above: Workers inspect honeycombs at a commercial honey farm.

Above: Apple blossom flowers early in the year, thus the first flow of the year may be apple blossom honey.

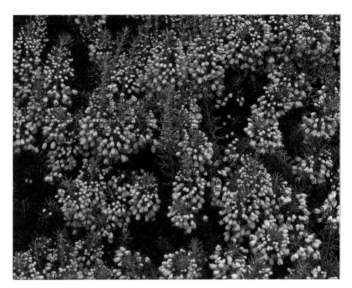

Above: Heather generally flowers toward the end of the year, and there will possibly be a late flow of honey if the weather conditions are clement.

combs with honey, or less than one full box, the bees will need this for themselves, especially if winter is approaching. If there are at least two or three boxes on each of your hives, then it may be worth extracting.

Late spring honey harvest

Perhaps your hives are located in an area with large patches of spring wildflowers or apple orchards. By the end of spring, the hives may have several boxes of wildflower or apple blossom honey which you could extract, thinking that late spring or summer crops will provide a second, more extensive harvest, or that autumn-flowering heather could provide a late flow. As long as you are sure that,

given average climatic conditions, these later flows will occur, the early honey can safely be removed. However, if the later flows fail owing to lack of rain or some other unexpected weather pattern, you will end up having to feed your bees over the winter. Also, between the spring flows and the summer flows there is often a lengthy gap of some weeks, during which your bees could starve if you take away their honey.

SUFFICIENT FOOD

Always ensure that if you do decide to take the honey, at whatever time of year, you leave a super of honey immediately above the brood box and ensure that it is full of frames of honey. Move frames from some of the other boxes if required.

Above: If the later honey flow fails, it may become necessary to feed the bees to prevent them starving when there is a dearth.

REMOVING THE HONEY

Before you begin harvesting, plan exactly how and where you will extract the honey, and ensure you have the right equipment to hand, to make the operation more efficient.

When you know that the honey is ready to be harvested, it is the right time to use your field equipment and remove the boxes of honey from the hives to your extraction area. Before you start, gently smoke the hive you are working on to calm the bees – remember, you are now robbing them and they may

not be happy about it! Keep the smoker handy and alight to top up with a few puffs of smoke if necessary.

From your pre-harvest inspection, you will have noticed that there will be supers that are full of honey frames and each frame has a full load of capped honey; supers that have some frames

that are full or mostly full of honey (harvestable); and others that have insufficient honey or capped honey to extract. (Leave those with less than 75% capped honey.)

After using either bee escape boards or fume boards to clear the bees from the honey supers, you can remove any supers of full honey frames on to your barrow or carrier and take them to the house or enclosed vehicle. Do this quickly, because the bees will soon start trying to get their honey back.

Dealing with mixed frames

Supers containing a mix of full frames with capped honey and frames that are not suitable for harvesting must be dealt with individually as follows: Place an empty super on an upturned lid on the ground. Keep a cover handy. Remove and inspect each frame of honey from the honey super. (As you go through the frames, you may have to brush off any remaining bees with your bee brush).

Place those frames that are full of capped honey into the empty super on the ground, covering this box with a lid to help keep the bees away.

Repeat this process until you have a box(es) of full honey frames ready to move to your extraction facility, as well as boxes still on the hive that have an assortment of honey frames that are not yet suitable for harvesting.

Leave those frames that are not adequate for harvesting in the hive. When the receiving box is full of harvestable frames, cover it up so that it is bee-proof.

Above: Before removing the frames full of capped honey from the supers, gently puff some smoke into the hive to calm the bees, before taking their stores of food.

Above: A beekeeper removes a honey super from a hive prior to extraction.

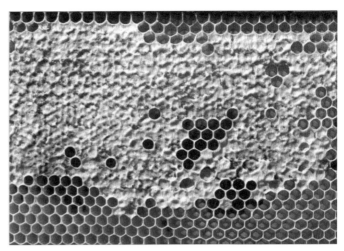

Above: Close up, the frames of comb show some empty cells and other cells that are full of honey, with wax cappings covering them to protect the contents.

Place all these non-suitable frames together in the hive and fill any gaps with spare empty frames of comb or wax foundation. Close up and move onto the next hive repeating the process. Note that when all the frames containing honey have been removed, re-assemble the hive, ensuring that the bees have at least one super of full or partially full honey combs above the queen excluder, and then transfer the boxes containing the removed honey frames to the house or shed, using your vehicle or wheelbarrow, so that you can get ready to start the extraction of the honey.

PREVENT ROBBING

Robbing can build up very rapidly. No matter how careful you are, honey will inevitably drip all over the place, attracting robbing, so the sooner you get it safely back home, the better. Keep the supers covered as you collect them.

Above: Honey bees congregate on a new frame that has been put into the hive to replace those that have been removed for extraction.

Above: Frames of capped honey are placed into an empty super, then covered.

EXTRACTING THE HONEY

Once all the frames full of honey have been taken from the hives and brought into your bee-proof room, it is time to start extracting the harvest.

You now have a stack of honey supers sitting on trays or newspapers in the kitchen or extraction shed, and you can start the exciting process of extracting honey.

First, place an empty honey super next to the stack of full supers. Take the first frame from the first honey super and hold it over your uncapping container. Heat up an electric knife for uncapping and run it down the comb, removing the cell cappings, which will fall into the container, or use a knife that you have placed in hot water. Scrape off the cappings from any uneven parts of the comb and repeat the process on the other side of the comb. When the uncapping process is complete, place the frame in the extractor.

Above: This shallow frame is from a National hive. It is full of honey and covered in wax. The wax is removed, or uncapped, and the honey is then extracted.

Above: A stack of honey supers sits on newspapers awaiting extraction.

Above: Before extraction of honey, each frame must have the wax removed.

Above: Using a hot knife is the quickest way to remove the wax cappings.

Above: Here the wax is removed using a metal trowel.

Above: The wax cappings on uneven comb can be removed with an uncapping fork.

Above: In your bee-proof room, a frame with honey is placed in the extractor.

Repeat this uncapping process for other frames until you have sufficient frames ready to start extracting. This is usually two or four frames for a hobby extractor. Try to ensure that the frames in the extractor are of even weight, and place it on a raised platform to make decanting the honey easier once the extractor is full.

Using the extractor

Now you can start the extractor. If it is a hand-driven model, wind it up gently to full speed and keep going until you see little or no more honey flying out of the cells and hitting the side of the extractor container. If it is motor-driven, slowly turn it up to full speed. Keep an eye out for large chunks of comb flying off the frames, as this could mean that you are going too fast and breaking the combs.

If you are using a tangential extractor, spin the frames until honey stops flying out. Once the extractor has stopped turning, remove each

frame and turn it around so that the other side is facing outward, and carry out another full spin.

Even though this is a more laborious process, tangential extractors do tend to extract more honey than the radial varieties, especially if you are spinning by hand. The main advantage of radial

Above: If you harvest comb honey, which is the wax and the honey, it saves a lot of time. Both the wax and honey are edible.

extractors is that you don't have to stop and swap the frames around at the halfway stage.

Once all the honey has been extracted from your frames, remove them from the extractor and place them in the empty super standing next to the stacks. Make sure this super of empty frames also has newspapers below it, because even when they are empty, the frames will drip. When you think that you have extracted the honey from all the frames in your first super, this super becomes the new empty box to receive the next frames.

PRODUCTION LINE

Extracting the honey can be a lengthy business if you go it alone. However, if you can persuade a friend or family member to help by uncapping the combs while you spin the extractor, the process can be speeded up quite considerably.

AFTER THE HARVEST

Once the honey harvest is nearly over, it is time to think about storing the honey, and then cleaning the equipment and tidying it away for future use.

Preparing to empty the extractor

If all goes well, your extractor will very soon need emptying of its delicious contents. Unless you have a pump that can handle viscous liquids, the extractor should be set on a platform above your storage buckets before spinning, as mentioned previously. However, if this is not possible in your bee-proof room, you should be able to lift it with a little help from a family member or friend. The smaller extractors suitable for hobby beekeepers do not hold a huge amount of honey, and therefore will not become too heavy to lift.

Transferring the honey

To remove the honey from the extractor, place the storage container either on the floor or on a stable stand, below the tap near the bottom of the extractor.

Hang the filter underneath the tap or spread it over the top of the storage container. Open the extractor tap and let the honey flow out through the filter into the storage container. If the filter becomes clogged, close the tap and stop the operation until it has been cleaned. When the extractor is fully drained, cover the storage container with a lid and repeat the extraction process with more frames.

Cleaning up

Once you have finished extracting honey, you will have many honey supers that are full of empty, wet frames. Before storing them away in your shed for the winter, you can place them out on the ground near the apiary for the bees to clean up. Some beekeepers claim that this can cause an outbreak of robbing,

but there is little evidence to suggest that this will happen in an apiary of strong hives. Once they are cleaned of all remaining traces of honey, stack them in a shed, first lightly spraying the frames with *Bacillus thuringiensis*, which is available from bee supply stores. This will protect the combs from wax moths and will not contaminate the wax. In areas where winter temperatures are cold enough to freeze your frames, there will be no need for any further precautions.

Honey storage

You should now have a container or two full of beautiful liquid honey. The ideal container also has a tap, like the extractor. Most honeys will set sooner or later – indeed, some set very rapidly, such as rapeseed honey – and it is a good idea to decant your honey into

Above: After extraction, the honey is filtered before bottling.

Above: Transfer the filtered honey into clean jars and seal immediately, label with the date and type of honey, then store the honey in a dark area.

Above: Once the honey has been completely removed from the frames, they are sprayed with Bacillus thuringiensis, which protects against wax moths while in storage.

smaller jars before it sets, and seal them well. However, to avoid bubbles of air, leave for one day before bottling.

Honey will not deteriorate in storage as long as moisture cannot enter the containers. If there is too much water content in your honey, for example if you extracted the combs before they were three-quarters capped, or if you allowed moisture to enter the container after extraction, the honey may ferment and will burst the containers. This honey should be pasteurized. A honey refractometer, which you may be able to borrow or hire, tests for the moisture content of the honey. If it is below 17 percent, you should have no problems. However, this figure is higher for some honeys, such as heather honey, and it also depends on the wild yeast content.

Harvesting the wax

Drain the honey from the cappings for a day or two. Wash the cappings in water, then melt them in a double boiler. Strain through muslin to remove any foreign bodies, then pour the wax into a mould.

Above: The wax cappings are strained through muslin to remove pollen or unwanted foreign bodies, and the honey is collected in a bowl.

Above: The wax cappings can be melted and used to make candles.

THE USES OF HONEY

There are many ways honey can be used – as a sweetener, a nutritious food, a base for making drinks and a natural medicine, to mention only a few.

Above: The fruits of the harvest are chunky honeycomb and jars of golden honey. The taste is dependent on the diversity of plants from which the bees collect the nectar.

BAKING WITH HONEY

When using honey to make sweet confections, you will probably need to make some adjustments to the recipe.

For cakes and biscuits, replace up to half the sugar with honey, and reduce the liquid in the mixture by 50ml/2fl oz/¼ cup for every 200g/7oz/1 cup of sugar replaced. For fruit bars, replace up to ²/₃ of the sugar with honey, again reducing the liquid by 50ml/2fl oz/¼ cup for every 200g/7oz/1 cup of sugar that is replaced.

Reduce the oven temperature from 180°C/350°F/Gas 4 to 170°C/325°F/Gas 3.

As well as being a flavourful ingredient in cooking, honey is used in drinks, hand and body lotions, medicines and soaps.

Honey in the kitchen

Always make sure that you have honey available for cooking. It is one of the most versatile ingredients for sweet and savoury dishes. Honey is delicious spread on toast, as a topping for ice cream, stirred into yoghurt or as an ingredient in baking sweet cakes, puddings and biscuits (cookies). It can also be used in glazes and marinades for fish and poultry, and to add flavouring to savoury meat and fish dishes.

To make good mead (a fermented honey drink dating back centuries) is surprisingly difficult and requires time and dedication, but many people would say it is worth the wait. Honey beer is another favourite drink, and it is easier and quicker to make than mead.

Above: For breakfast, toast and honey is an energizing start to the day.

Above: Classic honey and spice buns will be popular as a teatime treat.

Above: Pears poached in scented honey make a delectable dessert.

Above: Mead, or honey wine, is an alcoholic drink made from honey that has been fermented.

Honey as a medicine

Clinical research has shown that honey has a host of medicinal properties. It is now used in hospitals and clinics all over the world. For example, researchers in the USA found that a small dose of buckwheat honey given before bedtime

SELLING HONEY

If you are a small producer and wish to sell your honey, you should find plenty of buyers at country shows, farmers' markets, summer fairs and so on. Remember that putting your honey into small, fancy containers can give a much better financial return than selling it in large pots.

If you do sell your honey, remember to investigate and comply with all the government food production, selling and labelling laws.

provided better relief of night-time cough and insomnia in children than DextroMethorphan (DM), a cough suppressant found in many over-the-counter cold medications. When mixed with herbs, honey is believed to relieve stress and digestive complaints.

Honey is an ideal topical treatment for sunburn, burns and minor abrasions. It seals off the area from foreign substances and fights infections.

Manuka honey from New Zealand even has proven anti-bacterial properties, and is used successfully in wound and burn treatments.

Honey for beauty

Because of its emollient qualities, honey is often used in health and beauty aids.

Mouth wash: Try a tablespoon of honey in a glass of warm water. It cleans teeth and dentures, kills germs and leaves the breath smelling sweet.

For shiny hair: Hair shiner based on honey can be applied after washing, or try a small amount of neat honey instead.

Skin toner: Honey is used in facial masks, lotions and cleansers, often with other natural ingredients. It cleanses and moisturizes, leaving the skin clean and supple, and many say it clears acne.

Above: Honey and beeswax are often used to make soaps and creams. As well as having an aromatic fragrance, these products are softening and soothing to the skin.

PROBLEMS AND DISEASES

Beekeepers, like all other livestock farmers, are presented with their share of problems, which can arise at any time of the year. Novice beekeepers may not be able to recognize some of these situations, and one of the reasons for regular hive inspections is to act quickly when problems arise. Livestock diseases can be very serious, and failure to act can even result in the law taking a hand and destroying your bees and hives. This may be avoided if you take advantage of expert advice available from your local beekeeping association. The following pages outline various problems and diseases, and will provide you with possible solutions.

Left: Recognizing what a healthy hive looks and smells like will enable you to identify when something is amiss with your bees.

QUEEN PROBLEMS

The queen is the chief production unit of the hive, and as the mother of all the bees in the hive, the entire temper and pace of the colony comes from her.

A colony with a good queen will have eggs, young brood and capped brood, and if seen, she will be moving at a steady pace around the comb. Any problems with the queen can result in bad-tempered bees, low honey production or even the failure and death of the entire colony. It is important that you are able to recognize these problems and deal with them in a timely manner, before disaster strikes.

Above: An open queen cup indicates that a new queen has hatched.

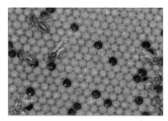

Above: A healthy brood, with most cells filled, is the ideal situation in the hive.

Take care with the queen

It is worth remembering that many queen problems are caused by the beekeeper. Queens can be crushed, lost or injured, and so cease to lay eggs. When inspecting a hive, always be very careful that you don't injure or kill the queen. Other problems that might be caused by the beekeeper are: if he or she introduces a new queen while the old queen is still present; if

Right: A beekeeper checks for the presence of the queen.

Below: The queen can usually be seen surrounded by attendants.

PROBLEM	SOLUTION/ACTION
No brood present in any of the cells.	The queen has probably died or failed, and a new one should be introduced or the colony united with another colony.
Sealed brood but no eggs.	The colony may have swarmed. Check in three weeks for the presence of eggs, or look for a virgin queen, which can be difficult to spot.
Drone brood only, with one egg in each cell.	A drone-laying queen. Re-queen or unite the colony after killing the drone layer.
Drone brood only.	This is often seen in worker cells, with the eggs not reaching the cell base, and means your queen is dead, and laying workers are present. See treatment later in this book.
Drone brood only with several eggs in worker cells.	Laying workers are present. In the absence of a queen they lay unfertilized eggs.
No brood, and small, excitable queen on the comb.	A virgin queen is present and probably not yet mated. Check for eggs in a week.
A very spotty brood pattern, with capped brood but lots of empty cells.	There may be a failing queen, or other causes include disease, pesticide poisoning or insufficient room to store honey during a heavy flow. Seek advice.
A small but good brood pattern.	A newly mated queen, which has just started laying. Keep checking to make sure of a build-up of brood.

the beekeeper introduces a new queen after the hive has been taken over by laying workers; the beekeeper fails to find the queen and then makes wrong assumptions about the state of the colony.

Solving queen problems

The table above is a guide to identifying and solving the most common queen problems. They can occur at any time and are not always easy for beginners to identify, but if you get the help of an experienced beekeeper, they will be able to spot them and find a solution. If you have a failing queen or a drone-laying queen, the colony will die. While she lives you will not be able to introduce a new, healthy queen, so you must kill her.

A NEW QUEEN AND LAYING WORKERS

There are many occasions when a beekeeper has to intervene in the natural process and introduce a new queen to the hive, or remove laying workers.

The introduction of a new queen may be necessary for many reasons. The old queen may have died or she may have become a drone layer, causing a shortage of workers. If the old queen's rate of egg laying has declined dramatically and, owing to lack of numbers, the colony is subsequently not thriving or taking advantage of honey flows, a new queen may be a necessity. Sometimes, the colony may become bad-tempered and a new, gentle queen will usually correct this problem.

Regular re-queening

Some beekeepers prefer to re-queen annually or bi-annually to maintain a high egg-laying rate and a high honey production rate. If you want to change the race of bee that you have in your apiary, you will also need a new queen. Have a new queen ready to introduce into the hive, and make sure the old queen is removed a day or two before you introduce your new queen. If you don't do this, the new queen will probably be rejected by the colony or the two queens will fight, and you can be sure that the old queen will win. The new queen should be in an introduction cage with a couple of

Above: The queen, toward the centre, is from New Zealand. These queens are reputed to be gentle and quiet, and are generally disinclined to swarm.

Above: Several drone cells in the hive can signal that the queen needs replacing.

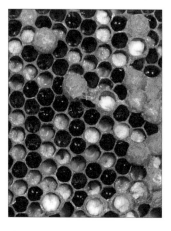

Above: Many drone larvae are shown in the largest cells in the hive.

Above: Drones start to emerge from their cells.

Above: Queen cages are shown holding a queen with attendant worker bees.

Above: The cage is opened in a plastic bag so the workers can escape.

Above: The cage is inserted between two frames; the bees will feed the queen.

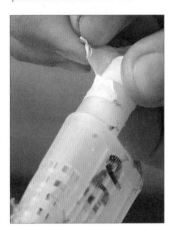

Above: Remove the plastic cap from the candy when you put the cage in the hive.

Above: The bees will eat the candy and within a day, the queen will be released.

BUYING QUEEN BEES

Unless you have reared your own queens, you can buy them from any reputable queen rearer. The new queen will arrive in the post in a carton – queens are posted all over the world by this method. If you are not ready to insert the queen, keep her in a warm place in her cage for a few days. Place drops of water around the cage.

attendants. There will be a candy plug at the cage entrance, and over this, a plastic cover. Remove the plastic cover and place the new queen in her introduction cage in the hive between two frames of brood. Make sure that the cage entrance is facing slightly upward and that it is free of any obstructions.

Within three days, check the hive to ensure that the queen has left the cage and is free on a comb. A few days later, check for eggs. If you find a good laying pattern, the introduction has gone well.

Dealing with laying workers

If a queen fails and the hive has no hope of raising another queen, some of the workers may lay sterile eggs, in competition with each other, and there are often several eggs per cell. The result is small, useless drones and the colony is doomed. Any attempt to re-queen the colony will fail because the new queen will be killed. Never try to unite a hive of laying workers with a healthy hive. If the colony is small it should be disbanded, but if it is still large, try the following:

Move the colony about 200 m (220 yds) from its original position, then tip all the bees out of the hive.

Take the hive back to its original position. The laying workers should be unable to find their way back to the hive. Next, re-queen the hive.

This method can be successful, but it does depend on all the laying workers being removed at a distance well away from the hive. The workers should become disorientated and unable to return.

AGGRESSIVE COLONIES AND ROBBING

Some bees can be plain nasty – they will easily be roused into an aggressive state by the beekeeper trying to inspect them and will attack anyone nearby.

When colonies show aggression, it is usually owing to one of the following reasons: it could be the genetic make-up of the particular strain of bees; animal interference with the hive, for instance skunks in the US; a difficult position, for example under power lines or next to a busy road; the collection of venom from the hive using an electric grid, in which the bees are given an electric shock to stimulate them to sting, so the the venom can be collected; or too much interference from the beekeeper.

An aggressive colony will cause quieter colonies to get worked up as well, until the whole apiary is in a state of uproar. There is no point in keeping colonies like this, and re-queening with a gentle queen is usually the answer. However, before

Below: Bees behave angrily if they have been interfered with by animals or by the beekeeper, or they may just be an aggressive strain.

Above: When bees become agitated, they may all react together and mob the beekeeper. Sometimes this behaviour occurs when robbing has taken place.

re-queening, make sure that the problem cannot be corrected by moving the hive or protecting it from animals.

Inspecting an aggressive hive

This can be a daunting prospect for beekeepers, however experienced. A good operational plan is needed for this manipulation and any inspection should be accompanied by a follow-up

re-queening expedition. On a warm evening, creep up to the aggressive colony and stuff a sponge strip into the entrance, thus sealing the hive. Move the hive 15–20m (50–65ft) from its previous position and open it up. Get help here if it is a heavy, multi-box affair – if you drop it you will be in a very unpleasant predicament. Leave the moved hive in place overnight.

Place an empty hive with some comb and frames of foundation on the original site. The next day, most of the stinging foragers will have gone to the dummy hive on the original site and you will be able to inspect the moved hive in relative peace.

Above: Reducing the entrance to the hive to just one bee space helps a colony being robbed to better defend itself and may deter invaders from entering the hive.

Above: Skunks are pests that rob hives and agitate the whole colony.

Below: Italian bees are more likely to rob hives than other strains of bees.

Kill the bad queen, and the next day introduce a new, gentle queen. Now introduce another gentle queen to the hive containing the stinging foragers.

Robbing

The robbing of hives by bees from other hives can be common when there is no honey flow. There are a few causes that make a hive vulnerable. It could be that a small nucleus hive has been placed in the apiary without reducing the entrance to one bee space so that the bees inside can defend themselves, or a hive with stores is weakened by disease or some other problem and is unable to defend itself. The beekeeper may have left frames of honeycomb in the apiary and gaps in the hives by not reassembling them correctly after inspection. Another reason is simple genetics – some races of bee are more prone to robbing than others, and the Italian bee in particular will take advantage of any other hive weakness.

Preventing robbing is easier than stopping it once it has started. Avoid the circumstances listed above and keep an eye on the health of your colonies.

If robbing does start for some reason, you can attempt to limit any damage by ensuring that the robbed hive has just one very small entrance to reduce the chance of other animals entering. Use grass, wood or mud to block up all other gaps. If it is a nucleus hive, it is probably best to move the hive or hives somewhere out of flying range of the robbers. This method can also be used effectively with full-size hives.

UNITING COLONIES

Colonies may have to be united with other colonies for a variety of reasons, such as making a larger colony or dealing with a weak or queenless hive.

If you have created an artificial swarm as part of your swarm prevention measures, you may not want the extra colony. You can unite the new colony to a stronger one, and this may also help during a honey flow. Remember that one large colony produces more honey than two smaller ones. The other main reason to join two colonies is if you find a colony that is queenless toward the end of the year after harvesting. This colony needs to be united to another one for over-wintering.

Work out exactly why you are uniting two colonies before you begin, and ask yourself whether it is the most beneficial step to take. If, for example, you have a weak colony, why is it weak?

Above: Before uniting two colonies, the beekeeper looks for the queen. If two queens end up in the same hive, one will be killed, and it may be the best queen.

If you know it has a disease, you will just spread this to a healthy colony; if the weak colony has laying workers, indicating that there is no queen in that colony, the workers may kill the queen of the colony you are uniting it to.

Bee colonies do not unite naturally, and it is usual for colonies to fight each other rather than unite, so special precautions must be taken to ensure that the union proceeds as amicably as possible.

Once you have decided that uniting colonies is the right thing to do, you can safely go about it in one of two ways. The first gives the two colonies time to get to know each other, and the second confuses them right from the start.

Left: A swarm can be kept in a skep and placed in the hive at dusk.

Above: A sheet of newspaper, with holes in it is all that divides two colonies of bees.

Above: The united bees are transferred to another box.

Above: Once the bees have accepted the other colony, the boxes can be joined.

The newspaper method

This is safe for the bees, rarely fails and no special equipment is needed. Open up the hive that you want to unite to. Place a sheet of newspaper over the top of the box and if necessary tack it down to stop it blowing away. Make some slits in the paper with your hive tool. Lift the weaker hive off its floor and place it on the newspaper-covered box. Leave for a day or two and check if the two colonies have merged peacefully.

The confusion method

Bees are thought to recognize their nest mates primarily by odour, so if you can artificially mask this odour

Above: The beekeeper is spraying a solution of sugar into the hive, in an attempt to confuse the bees from two different hives and unite them peaceably.

KEEP ONLY ONE QUEEN

If both colonies that are being united have queens, first kill the queen of the weaker colony, or whichever queen you don't want. if you don't, the two queens will fight it out and it is possible that the weaker one will win.

with an alien scent even for a short time you will be able to temporarily confuse the bees.

Open up the stronger hive. Lift the weaker hive off its floor and bring it to the stronger hive. Quickly spray the top frames of the stronger hive with a non-toxic room freshener spray or a sugar syrup spray. Also spray the bottom frames of the weaker hive with the same spray and lift it on to the stronger hive.

Many beekeepers don't like using artificial sprays in the hive. Powdered sugar, sugar syrup spray and even flour will confuse them.

PESTICIDE POISONING

One of the worst hazards of beekeeping is poisoning by pesticide spray, but the problems can be resolved so that both beekeepers and farmers can go about their business in peace.

The use of pesticidal chemicals in modern farming arises from one of the great dilemmas in our agricultural systems: the need to protect our crops from insect pests on the one hand, and the requirement for other insects, such as bees, to visit those crops to pollinate them. Most horticulturalists and beekeepers recognize the problem, and fruit growers value bees for their vital contribution to pollination. Pesticide poisoning appears to be the result of ignorance, or simply lack of thought.

How pesticides affect bees

Chemical poisoning in bees affects the nervous system, leading to lack of co-ordination of the body functions. The alimentary system is also affected and this can lead to starvation.

Often, the pesticide is carried back to the hive in the form of dust or liquid and so it then affects the entire colony. Herbicides that are labelled 'safe for bees' can also be dangerous, because the chemical surfactants that are used to make them adhere to plant surfaces also make them stick to adult bees. These are then carried back to the hive, where they affect the brood.

Recognizing spray poisoning

The signs of poisoning from spray or dust application can be recognized as follows: there will be large numbers of dead bees at the hive entrance. The proboscis of the dead bees will be extended, and there will be crawling, trembling or aggressive bees around the outside of the hive. Bees will be refused admission to the hive.

Above: If you find huge numbers of bees dead in front of the hive, they have almost certainly fallen victim to pesticide poisoning. The use of pesticides is higher in rural areas where food is grown; urban areas use much smaller quantities.

Above: If you know there is pesticide spraying about to occur, take measures to close up the hive thus minimizing the impact of the chemical.

Above: Put a gauze-lined lid on top of the hive and cover the whole with a dampened sheet to discourage the bees from flying in and out.

Most farmers will help beekeepers in this way. You can also register with your local beekeeping association 'spray liaison' scheme. The co-ordinator of the scheme will let you know of future spraying in your area. If spraying must go ahead, move the colonies at least three miles from the spray zone, but make sure that you are not just moving them into another spray zone that you haven't been informed about.

If you are unable to move the colonies to another area, carry out the following preventive measures on each hive before the spraying starts: remove the hive lid and place an extra empty box on top of the hive. Place a wet sponge in this box on top of the frames. Tack some gauze over the box so that no bees can escape.

Replace the hive lid, but raise it on wooden slats to allow ventilation. Then block up the hive entrance with gauze or close the hive.

In this way, the bees will not be able to leave the hive for the period of spraying, but with sufficient water and some ventilation, they will survive until the spraying is over.

Be careful of your diagnosis, because some of these symptoms can also be caused by other problems, such as starvation. Record as many details as possible, such as the number of dead bees per colony, the colour of pollen on the dead bees and the weather conditions. Take large samples of the dead bees (at least three samples of 300 bees) and send them to your bee disease laboratory. Photographs may help.

Preventive measures

Always let the local farmers know that you are a beekeeper and ask them to warn you if they are going to spray.

Above: Before covering and sealing off the hive to prevent pesticide poisoning, the beekeeper must ensure that there is adequate ventilation and water.

Above: To avoid spraying, another option is to transport the hives to a safer area.

IDENTIFYING AND PREVENTING DISEASE

As with any other livestock, honey bees suffer from a range of diseases and it is the responsibility of the beekeeper to recognize their symptoms early on and know how to act.

It is very important to find out as much as you can about bee diseases, because if you let disease take hold in your apiary, it will not only destroy your own colonies but also spread rapidly to your neighbours' colonies. If you neglect this aspect of livestock management, a visit from a beekeeping disease inspector could result in all of your hives being destroyed.

The subject of bee diseases is huge. While you can learn much about them in this book, it is worth looking out for any new developments in treatment and diagnosis by reading a good beekeeping magazine.

Vigilance, regular inspections and early preventive action will help to keep your colonies strong.

Above: Many dead bees may indicate disease, poisoning, or the bees in the hive have just cleared out dead ones. Whichever it is, it requires investigation.

Above: In some countries, bees are transported from place to place so that they can pollinate crops of fruit and nuts, as in this orchard (above). This can cause environmental stress and lowered immune systems, which makes them susceptible to disease.

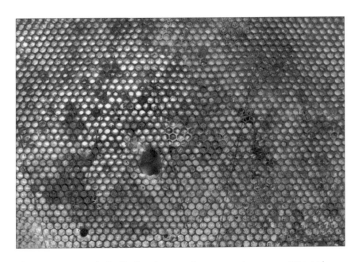

Above: Damage seen in the hive is owing to an infestation of wax moths.

Taking action

Beekeepers can get into trouble through inaction, and unfortunately many do trust to luck in their disease management. Often their first real proof of a problem is a hive full of wax moths, which take over any failing colony that can no longer defend itself.

If you think there may be a problem in the hive, do not be afraid to ask a beekeeper with more experience, a local bee disease inspector or apicultural extension specialist. The problem will not just go away on its own.

There are two main types of bee disease: those affecting the brood, and those affecting adult bees. There is also a pest that is an indicator of disease, the wax moth. The next few pages in this book will describe bee diseases and offer possible solutions.

Developing a strategy

The best way to avoid problems is to develop what is known as an Integrated Pest Management (IPM) strategy. This really is the only systematic way to manage pests and diseases.

Its components are: prevention and awareness, observation and monitoring, and prompt Intervention when required.

Preventing disease

Preventing disease in bees is generally a matter for the beekeeper. Finding a sick colony and taking them to the vet is not possible. You have to sort it out yourself, and if you are very inexperienced you should know when to ask for help from another beekeeper, your local association or a bee disease inspector.

Many disease-causing organisms may be present in background quantities in beehives, and these are usually kept in check by the bees' immune systems unless the colony is subjected to environmental stress, such as being repeatedly moved or subjected to damp, cold conditions. This may be one of the main causes of declining bee numbers. Keep your hives healthy and strong with good management.

Notification of diseases

You must find out which diseases require notification to your local beekeeping authority. In most countries, for example, American Foul Brood (AFB) has to be declared – it is an extremely dangerous disease. New pests and diseases found in your hive must also be declared; for example *Tropilaelaps clarae* (a Varroa-like mite) must be notified in the UK.

Above: Wax moths have taken over a weak colony of bees, laying their eggs and causing quite a lot of damage. A wax moth caterpillar is present here.

WAX MOTH AND AMERICAN FOUL BROOD

The presence of wax moths is a good indicator of a colony in trouble, and since brood diseases are highly infectious, they both should be caught and dealt with at an early stage.

Wax moths are specially adapted insects that depend on bee colonies and beehives for their existence. They lay their eggs in the hive, and the moth larvae move through the comb, eating honey, pollen and beeswax. A strong, healthy colony can cope with them, whereas weak or diseased colonies unable to defend themselves will usually be overrun with wax moths very quickly.

Their purpose in nature is to destroy weak and diseased colonies and prevent them from becoming a reservoir of disease. If you see any damage by wax moth, check the hive out thoroughly for disease or queenlessness. You may be able to save it by early intervention.

Identifying wax moths

There are two types of wax moth, the Greater Wax Moth (*Galleria mellonella*), a dull grey in colour, and the Lesser Wax

Moth (*Achroia grisella*), silvery to white. They are easy to identify since they are the only moths you will see in the hive.

Preventing wax moth damage

Healthy colonies keep wax moth damage in check, so you should ensure that your colonies are strong. Wax moths will eat

and destroy stored comb, so watch out for them in the shed over winter. Stagger the boxes of stored comb so that light can enter the empty hives, and very lightly spray each frame with *Bacillus thuringiensis* (obtainable from any bee supply store). Freezing temperatures will destroy all stages of the wax moth.

Above: The death's head hawk moth raids beehives for their honey, but unlike the wax moth, does not take over the hive.

Above: The greater wax moth (above) and the lesser wax moth (top) prey on bees, eating comb and pollen, and laying their eggs in the hive. They cause great damage and will overrun a weakened colony.

American Foul Brood (AFB)

The most serious brood disease, AFB, is caused by the spore-forming bacterium *Paenibacillus larvae*. Young bee larvae under 24 hours old are most at risk from ingesting AFB spores in their food, and will die after their cells are sealed. The spores are very infectious and will move from hive to hive during outbreaks of robbing, or are carried by drone bees

Below: To prevent wax moth larvae feeding off debris and wax in the comb over winter, the combs are sprayed with a solution containing Bacillus thuringiensis.

entering other hives. Commonly, they are spread by the beekeeper, carried on the hive tool and gloves. Identification is difficult in the early stages, but look for the 'pepperpot' look of filled and empty cells; brown, often sunken and sometimes pierced cell cappings; brown, discoloured remains of the larvae in the cell, which can be drawn out like a rope with a matchstick; open cells exhibiting the pupal tongue of the dead larva: a foul smell in the cells and frames.

To treat AFB, close up the hive, wash all your clothing and equipment and call the local bee disease inspector

immediately. If you leave this disease unchecked, it will destroy your colony and spread to other colonies in the area. The bee disease inspector will either treat it with the bacteriostat oxytetracycline or, more commonly, by burning the hive, equipment and bees, depending on how far the disease has progressed. Once you have AFB, the spores will remain in the equipment for a long time.

Increased treatment of colonies with oxytetracycline has led to bacterial resistance and further problems, whereas the more effective policy of burning infected hives in the UK and New Zealand has led to a marked decline in American Foul Brood in these two countries.

BEEKEEPING ASSOCIATIONS

A good reason for belonging to your local beekeeping group is that you can ask for expert advice if you notice anything amiss in your hives. An unusual smell in the hive may indicate that there is a problem that should be dealt with.

Above: These bees died after the colony was weakened by American Foul Brood.

Above: A pupa can be drawn out like a rope, indicating American Foul Brood.

Above: The frames and hives affected by American Foul Brood are burned.

OTHER BROOD DISEASES

While AFB is the most virulent and devastating of brood diseases, there are others to watch out for, although with care and immediate treatment, the hive can often be saved.

Certain diseases that affect the brood cannot be treated and just run their course; others require early treatment.

European Foul Brood (EFB)

This disease is caused by the non-spore-forming bacterium, *Melissococcus pluton*. The bacterium infests the gut of the larva and causes starvation. If caught in its early stage, the hive may be saved, and occasionally if a honey flow starts, the problem can disappear. It is a notifiable disease in many countries.

There are several indications of EFB: the larvae turn an off-white colour, adopt unnatural positions and lose definition. Infected cells are not usually capped; if they are capped, the cell cappings will appear to be sunken and are often perforated. There will be a typical 'foul brood' smell that will be obvious as soon as you open the hive. EFB is often referred to as a stress-related disease. The removal of any type of

Above: Listening to the hive and noting any foul smell may mean that you take action quickly enough to effect a difference to the hive, and build up a healthy, strong colony.

stress, such as moving the hives, or the onset of a honey flow, may relieve the situation. The bee disease inspector may also use oxytetracycline if appropriate.

Sacbrood

This is a virus disease for which there is no treatment currently, although the beekeeper can remove the affected larvae in an attempt to bring the outbreak to a halt. Larvae infected with sacbrood die in their sealed cells. They become light yellow in colour, with tough skins, then the skin darkens and the outer layer becomes loose, forming a sac enclosing a watery fluid. The brood lies stretched out lengthwise in the sealed cell. After the death of the insect, the cell is partly or fully opened, and the worker bees remove it from the hive.

The virus is spread in the nest by the house bees evacuating the dead brood. However, this virus does not survive long, and the disease may disappear during the honey-flow period. Serious outbreaks are not common, and usually no action is necessary. If control is needed, then the colony must be re-queened to introduce a different genetic mix.

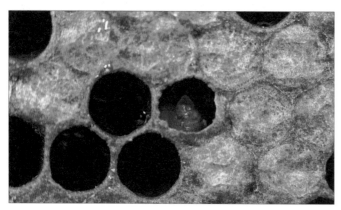

Above: In sacbrood disease, the outer layer of the larvae becomes loose and filled with a watery fluid, and the larvae die in their cells.

Above: A closer view of mummified chalkbrood larvae.

Above: The open cells reveal the presence of sacbrood in this comb.

Above: At the Bee Research Laboratory in Maryland, USA, a technician prepares to examine bees to determine whether the cause of death is due to parasites or disease.

Chalkbrood

This is a fungal disease caused by *Ascosphaera apis* that affects both sealed and unsealed brood. The incidence and spread of chalkbrood infection is difficult to predict. Colonies under stress from shortage of space, food or water, poor weather or other infections seem to be more susceptible to chalkbrood infection. The larvae usually die within two days of cell capping.

Fungal spores in brood food ingested by the bee larva germinate in the hindgut, or by growth through the cuticle. When the cells are capped, fungal mycelia develop from the spores and take over the larva, giving a noticeable 'cotton-wool' appearance. The larva shrinks and dries to a white or grey-black chalk-like mummy, the colour depending on which sexual type of *A. apis* has invaded. Bees remove these dried hexagonal mummies and often deposit them on the alighting board – a sure indication of an outbreak.

There is no chemical treatment. Generally the most effective policy is to keep hives strong, healthy and free from stress. Research also shows that good ventilation of the hive may help.

Stonebrood

This disease is caused by a fungus belonging to the genus *Aspergillus*. It attacks the brood and transforms the larva into a hard, stone-like mummified object which is found lying in open cells.

Adult bees may also be attacked and they too are killed in the process. There is no recommended treatment, but the bees remove the dead brood, and the colony usually recovers in a short space of time.

Above: Sometimes the first sign of chalkbrood is noticing hard mummified bees lying around on the alighting board, after they have been removed from the hive.

ADULT BEE DISEASES

Diseases in adult bees can be as confusing to the untrained eye as brood diseases and can be just as deadly if left unattended, so always take prompt action.

Many problems and syndromes of adult bees, rather than those affecting larvae, appear to be linked to various disease-causing organisms. It is important to get good advice from an experienced beekeeper or bee inspector if you are unsure whether anything is wrong.

Parasitic diseases

There are several parasites that can affect bees, all of them can cause widespread problems in a colony. *Nosema apis* is a small unicellular parasite, now considered to be a fungus, that causes widespread disease in honey bees. If the spores are eaten by bees, they germinate and invade the gut wall, where they multiply then pass through the waste. Infected bees don't live as long as healthy bees, and queens suffer damage to their reproductive organs. Bee populations can drop dramatically if infected with *N. apis*. Spring tends to be a high-risk time, until sick bees die off and new healthy brood emerges. When parasite levels are low, the colony appears normal.

Nosema ceranae is a parasite that may have jumped from the Asiatic honey bee, *Apis cerana*, to the European honey bee. European hives suffered enormous colony losses from around 2004, and it was at this time that *N. ceranae*, previously thought to be confined to Eastern hives, was discovered in Europe.

The sequence of events led beekeepers and scientists to link the appearance of *N. ceranae* with Colony Collapse Disorder, a massive

Above: Dead bees are examined in a bee laboratory to see if they carry the spores of a Nosema *parasite.*

and sudden die-off of honey bees. This disease can cause severe and rapid loss of colonies. There are no external signs of *Nosema* except a dwindling of the colony and, in extreme cases, the total disappearance of all of the adult bees. Dysentery can help to spread *Nosema*, but it is not a symptom of the disease. Some countries permit the use of a chemical, Fumidil B (fumagillin), which can be used in bee sugar syrup feed. Research is ongoing as to its effectiveness and again, the best

preventive treatment is good hive management, with the aim of producing strong healthy colonies.

Dysentery

This disease is a symptom of a gut problem – probably excessive water accumulation in the rectum. This can be caused by bad feed or other nutritional factors, and can be recognized by a large amount of fecal spotting on the hive.

Virus diseases

These have various exotic names such as black queen cell virus, filamentous virus and bee virus Y, and can show several symptoms in bees, most of which will

Above: A beekeeper in Europe holds dead honey bees (Apis mellifera), which have come from a hive that has been affected by Colony Collapse Disorder (CCD).

Colony Collapse Disorder (CCD)

This mysterious disease has affected hundreds of thousands of colonies worldwide. Although research is ongoing, there is no solution yet. Possible causes include malnutrition, climate change, pesticides, parasites, beekeeping practices and a limited gene pool.

The warning signs are that the hives affected contain pollen, honey and some emerging brood, but the adult bees have disappeared.

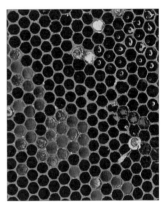

Above: The absence of adult bees in the hive is a typical symptom of CCD.

Below: The presence of laying workers indicates that the queen has died.

occur together. The signs include bees crawling around the hive entrance or on the alighting board in a semi-moribund state, and clearly unable to fly; bees with a black and greasy appearance; and bees with extended abdomens. These bees may be refused entry into the hive.

Bee viruses are known to be associated with parasitic diseases such as *Nosema apis* and perhaps *Nosema ceranae* infections may add to their harmful effects, possibly causing or contributing to Colony Collapse Disorder and Parasitic Mite Syndrome (PMS).

BEE MITES

Two bee mites particularly concern beekeepers at present: *Acarapis woodii* or the Acarine mite, and *Varroa destructor*, a mite that has caused huge problems for honey bees.

Keeping the colony strong and making regular checks is vital to prevent and combat mite infestations.

Acarapis woodii

The microscopic Acarine mite enters the bee's breathing apparatus (the tracheal system), multiplies there, and interferes with the bee's respiration. It derives its nourishment from the host's blood. The disease may not kill a whole colony in one year; the infestation can remain in a colony for several years. The mites alone cause little damage, but combined with other diseases or poor bee seasons, or both, caused by poor environmental conditions, these mites can eventually weaken the affected colony until it dies.

The Acarine mite is present in almost every beekeeping country in the world, although it is a particularly difficult problem in the USA. There are no certain methods for diagnosing the mite in the field, and it is difficult to detect an infestation early enough to take effective action. Current treatments rely on the use of menthol crystals and a proprietary treatment of oil of thyme in a gel base. Packets of these are usually placed in the hive, but advice on these treatments can be obtained from bee supply companies and from your bee disease inspector in areas where treatment is necessary. Treatment must end one month before the first nectar flow to avoid contaminating marketable honey and the wax.

Varroa destructor

This devastating mite jumped species from the Eastern honey bee, which can tolerate it, to the Western honey bee, which cannot. It has had a huge effect on beekeeping management. The mite pierces the bees' cuticle and sucks their blood. This action may aid the entry of other 'background' pathogens, and cause viral diseases. It further suggests a relationship with other syndromes such as CCD, which can cause the rapid loss of a colony through the disappearance of the worker bees and may also be associated with Nosema, and PMS, which causes a range of abnormal brood symptoms similar to AFB and EFB.

This is an essential part of the beekeeper's annual management plan. The treatment and control of *Varroa destructor* is a constantly evolving area of research, but essentially, a proprietary Varroa treatment of miticide strips or pastilles must be used in spring and autumn. These are readily available from bee supply companies and can be both chemical-based and organic. Breeding of certain hygienic and other traits into

Above: A Varroa *mite is attached to a bee, from which it will suck blood.*

Above: A parasitic mite from the order Acari is clinging to the hairs on a bee.

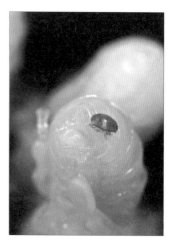

Above: Varroa mites are most easily seen on the larvae, indicating an infestation.

Above: A sticky board inserted into the hive catches and holds any mites that fall to the bottom.

Above: The beekeeper cuts up blocks of pastilles of an organic miticide against Varroa *to place around the hive.*

Above: Once the miticide pastilles are in place above the brood box, the queen excluder is replaced and the hive is closed.

bees is an ongoing process, and it may be that, in the future, Varroa-tolerant bees will be available.

A special impregnated sticky board can be placed in the hive to catch *Varroa* mites that drop down on it. Removing drone brood from the hive may also help, since the *Varroa* mites are attracted to drone larvae.

Find out about the latest treatments from your local association or bee disease inspector, and treat when other beekeepers in your area treat their bees. Concerted action in the neighbourhood really helps, and prevents re-invasion of mites from a neighbouring colony.

TROPILAELAPS CLARAE

Another potentially devastating mite is just around the corner. *Tropilaelaps clarae* exists in the Far East and is similar in effect but not in physical shape to *Varroa*. The mites are reddish brown and about 1 x 0.6mm. Beekeepers should look out for it in their hives.

Above: The impregnated strips with miticide are placed on top of the brood box; the beekeeper is placing the queen excluder on top to prevent the queen entering the supers.

OTHER BEE PESTS

Generally speaking, insect and animal predation on bees is not a major problem, and the effects tend to be localized, but there are a few predators that concentrate on beehives.

Depending on where you live in the world, your hives may be affected by various predators.

Small hive beetle

This destructive beast, *Aethina tumida*, has spread from Africa to the USA and as far as Australia, with devastating effect. The beetle and its larvae are adapted to living off honey-bee colonies. Larvae tunnel through comb with stored honey or pollen, damaging or destroying both cappings and comb. They defecate in honey, which becomes discoloured and slimy, and ferments. Heavy infestations cause bees to abscond; some beekeepers have reported the rapid collapse of even strong colonies. Treatments vary, but the most effective treatment is keeping strong, healthy colonies, coupled with minimizing the storage of empty frames of comb. Beetle traps are available that use non-toxic oil to suffocate the beetles.

Birds

Various types of birds may eat bees, but the main threat to bees from birds is the aptly named bee eater. Various species exist in many countries. Bee eaters tend to hunt bigger bees, and queens on their mating flights are an attractive target. Herons also pose a slight threat to bees as they can attack hives during very cold winters, but their overall effect is minimal. Most birds avoid bees and wasps.

Insects

There can be a problem with wasps, especially towards the end of the active season, when they will try to enter

Above: Various threats to the bees' wellbeing include bee eaters (top); hornets (centre left); ants (centre right); and infestation by mice (bottom).

beehives and take honey. Occasionally their presence can be overwhelming, but only small, weak colonies will succumb. Hornets, on the other hand, can be a deadly threat, especially the mandarin hornet, which can destroy a colony quickly. The Far Eastern honey bees can deal with the problem: they surround and 'ball' the hornet, raising the temperature to a degree above that which the hornet can withstand. Ants can be a nuisance if they occupy the hive.

Mice

These are a serious pest problem in bee colonies. Adult mice move into bee colonies in the autumn and make their nest in a corner of the hive away from the bee cluster. They chew comb and gnaw frames to make room for their nest, and their urine on the comb and the frames makes bees reluctant to clean out the nest in the spring. When the comb is repaired, the bees usually construct drone comb rather than worker-size cells. Mouse guards should be fitted before the weather cools.

Below: Bears can wreak havoc in an apiary; they may destroy every hive.

Skunks and raccoons

Night-time foraging animals may find bee colonies an easy food source. The skunk scratches on the outside of a hive, and as guard bees come out to investigate, they are eaten. Continued feeding means continued disturbance, and the bee colony may become very aggressive to the beekeeper. Hive entrance screening and elevation of the hive on a stand is usually an adequate deterrent.

Above: Raccoons can remove the lid from the hive to get at the honey inside. This can be prevented by placing a heavy stone or brick on the top.

Bears

Large animals such as bears can wreck apiaries. They can be a problem in the US and Europe. Bee stings have little effect on bears, and these protected animals can do great damage in a short time. Control of bears is difficult because once an apiary is found by bears, fencing may not deter them. Moving the apiary is one solution; or some beekeepers distract bears with old beehives, keeping them away from productive apiaries.

THE BEE GARDEN

One of the best ways you can help to slow down the decline in bees and other pollinating insects, apart from becoming a beekeeper, is to plant your own bee garden. By using the right plants and flowers you can create small, ecologically balanced areas that will attract not only honey bees but also bumblebees, butterflies and a host of other beneficial and often endangered insects. Go for the shrubs and flowers that are bee favourites, as described in the next few pages, and you will soon be enjoying the results.

Left: To benefit bees and other pollinating insects, you could seed an area of your garden with a variety of nectar-rich plants.

GARDEN PLANTS FOR BEES

There are many beautiful flowers and shrubs that are ideal for growing in the garden, and which also help bees by producing plenty of nectar at different seasons.

You will get most benefit once you attract insects, birds and other small creatures that feed on nectar-rich plants. Avoid showy flowers with large blooms – they may not contain any nectar. It is also worthwhile for beekeepers to learn about the trees, plants and crops in the countryside that provide good forage for bees. Some day, you may be looking for a new, bee-friendly site for your bees. Choose some of the following for your garden. Mahonia and wild rose are good early pollen sources, without which many bee colonies will not thrive.

Bee-friendly plants
- Bee balm (*Monarda didyma*)
- Common barberry (*Berberis vulgaris*)
- Cone flower (*Echinacea purpurea*)
- Daffodil (*Narcissus pseudonarcissus*)

- Heather (*Erica cinerea*)
- Hebe Neil's choice (*Hebe 'purple emperor'*)
- Hellebore (*Helleborus corsicus, H. foetidus*)
- Hyssop (*Hysoppus officinalis*)
- Lavender (*Lavandula stoechas*)
- Lemon balm (*Melissa officinalis*)
- Lungwort (*Pulmonaria*)
- Oregon grape (*Mahonia aquifolium*)
- Pussy willow (*Salix caprea*)
- Purple Rhododendron (*Rhododendron catawbiense*)
- Rosemary (*Rosemarinus officinalis*)
- Sage (*Salvia officinalis*)
- Snowdrop (*Galanthus nivalis*)
- Thrift (*Armeria maritima*)
- Viburnum (*Viburnum davidii*)
- Wallflower (*Erysimum cheiri*)
- Weigela (*Weigela florida*)

- Wild rose (*Rosa andersonii*)
- Winter-flowering honeysuckle (*Lonicera purpusii*)
- Zinnia (*Zinnia*)

This list is by no means exhaustive and you will no doubt find many other excellent plants for your area. Ask your local beekeeping association or your seed merchant. They can be a good source of knowledge and ideas.

Positioning the plants
Site as many plants as you can in full sun, in a sheltered place. Bees do not like to be blown around when trying to land on flowers and they prefer sun rather than shade. The plants that you use will depend on factors including the amount of shade or sun available, the soil type and acidity and the local

*Above: Bee balm (*Monarda didyma*). As its name implies, this plant attracts bees.*

*Above: Coneflowers (*Ecinachea purpurea*) attract many insects.*

*Above: Daffodils (*Narcissus pseudonarcissus*) are a source of nectar.*

*Above: Pink-flowering heather (*Erica cinerea) *flowers in summer and autumn.*

Above: Hebe 'purple emperor' *is known as the 'honey plant'.*

*Above: Hyssop (*Hyssopus officinalis). *This shrub attracts pollinating insects.*

climate. If you choose appropriate plants, you should gain increased nectar and pollen supplies for the bees.

Plants to avoid

Many popular flower varieties are hybridized for features that are valued by the gardener, such as disease resistance, flower size or colour, and bigger, longer blooms. Unfortunately much of this hybridization has reduced the production of nectar and pollen, and sometimes leaves the resulting plant completely sterile. This makes it useless to bees and other pollinators, so avoid these plants.

Hybrids are a cross between different species. A multiplication sign 'x' between the common parent genus name and the hybrid's species name, is used to show that a plant is a hybrid. Cultivars are plants of a single species bred to emphasize certain colours or sizes, and do not produce much nectar or pollen.

*Above: French lavender (*Lavandula stoechas) *is a popular plant with bees.*

*Above: Oregon grape (*Mahonia aquifolium).

*Above: Snowdrops (*Galanthus nivalis) *flower in the spring.*

*Above: Wallflowers (*Erysimum cheiri).

*Above: Wild rose (*Rosa andersonii) *is a good early pollen source.*

SPRING PLANTS

Spring-flowering plants give bees and other insects a really good start in foraging for nectar and pollen and help the bees to begin the production of honey after the winter.

In many country areas now, many of the following plants, especially the wild flowers, have been eradicated in order to make room for monocrop planting. However, there are still areas where some of these plants can be found so that the bees can make some surplus honey in the spring.

Good pollen sources are alder, crocus, dandelion, mahonia, poppy, wild rose and willow – make sure you look for these too, because pollen is a food for the bees, and just as necessary for bee colonies as nectar. Spring-flowering bulbs are an important source of early season pollen and nectar for honey bees and their new brood. There is little forage for them before the maythorn flowers or fruit-tree blossom opens, and this is another time when many colonies starve.

When thinking about the importance of bees, remember that one in three mouthfuls of our food is dependent on bee pollination. Crops such as apples, pears and plums depend up to 90 percent on bees for successful pollination.

- Alder (*Alnus glutinosa*)
- Common barberry (*Berberis vulgaris*)
- Black locust (*Robinia pseudoacacia*)
- Blackberry (*Rubus fruticosus*)
- Bluebell (*Hyacinthoides non-scripta*)
- Blueberry (*Hyacinthoides hispanica*)
- Blue gum eucalyptus (*Eucalyptus globulus*)
- Buckwheat (*Eriogonum douglasii*)
- Bugle (*Ajuga reptans*)
- Clover (*Trifolium pratense*)
- Cootamundra wattle (*Acacia baileyana*)
- Crab apple (*Malus sylvestris*)
- Crocus (*Crocus vernus*)

Above: Berberis Goldilocks.

- Currant (*Ribes spp.*)
- Daffodil (*Narcissus pseudonarcissus*)
- Dandelion (*Taraxacum officinale*)
- Flowering cherry (*Prunus serrulata*)
- Flowering currant (*Ribes sanguineum*)
- Forget-me-not (*Myosotis*)
- Gooseberry (*Ribes grossularia*)
- Gorse (*Ulex europaeus*)
- Hawthorn (*Crataegus monogyna*)
- Hazel (*Corylus avellana*)

*Above: Blackberry (*Rubus fruticosus*).*

*Above: Red clover (*Trifolium pratense*) is ornamental as well as rich in nectar.*

*Above: The bees get nectar and pollen from the bluebell (*Hyacinthoides hispanica*).*

*Above: Crab apple (*Malus evereste*) is a good pollinator for fruiting apples.*

Above: Redcurrants (Ribes sanguinem).

Above: Dandelions (Taraxacum officinale) are a good source of nectar.

Above: The flowers of the Higan cherry (Prunus x subhirtella) provide nectar.

Above: Hellebore (Helleborus) is another plant that foragers look for.

- Hellebore (*Helleborus foetidus*)
- Knapweed (*Centaurea nigra*)
- Lungwort (*Pulmonaria*)
- Oregon grape (*Mahonia aquifolium*)
- Maple (may flower in spring if warm)
- Pip fruits (apples/*malus*, pears/*pyrus*)
- Poppy (*Papaver rhoeas*)

- Raspberry (*Rubus idaeus*)
- Rosemary (*Rosemarinus officinalis*)
- Pussy willow (*Salix caprea*)
- Snowdrop (*Galanthus nivalis*)
- Stone fruits: cherry, plum, apricot, almond (*Prunus*)
- Sycamore (*Acer pseudoplatanus*)

- Thrift (*Armeria maritima*)
- Tulip tree (*Liriodendron tulipifera*)
- Tupelo (*Nyssa sylvatica*)
- Viburnum (*Viburnum davidii*)
- Wild rose (*Rosa andersonii*)
- Willow (*Salix caprea*)
- Wisteria (*Wisteria sinensis*)

Above: Common Poppy (Papaver rhoeas). L.

Above: Wisteria (Wisteria sinensis) is visited by many bees seeking nectar.

Above: Knapweed (Centaurea nigra) provides early summer nectar for bees.

Above: Bees can collect nectar from ornamental fruit trees in orchards.

EARLY SUMMER PLANTS

Your bees may have produced a surplus of honey from spring flowers, but there is often a gap between spring and summer flowers that you can fill with carefully chosen garden plants.

Take care that you do not begin the summer unprepared. In some areas there are few early summer flowerings, or even a complete gap of a few weeks between the spring blossoms and the abundance of summer. If you take all your spring honey off, leaving none in the hive, your bees will starve while they are waiting for the main flowering season to begin.

This is a time to make sure you have provided for the bees, by keeping the flowering period going in the garden and filling the gap between spring and high summer. You will have the added bonus of a garden full of colour for most of the year. Make sure that you plant the flowers in quite large patches to attract the bees. Once the summer is well under way, there should be no problem in providing plenty of nectar and pollen for the bees.

Flowers for bees in early summer

Plants that flower during the early summer gap are:

• Alfalfa/lucerne (*Medicago sativa*)
• Astilbe (*Astilbe ardensii sanguinea*)
• Barberry (*Berberis vulgaris*)
• Bellflower (*Campanula* spp.)
• Borage (*Borago officinalis*)
• Clover (*Trifolium pratense*)
• Columbine (*Aquilegia vulgaris*)
• Comfrey (*Symphytum officinale*)
• Common phacelia (*Phacelia distans*)
• Cotoneaster (*Cotoneaster horizontalis*)
• Cranesbill (*Geranium* spp.)
• Everlasting sweet pea (*Lathyrus latifolius*)

*Above: Columbine (*Aquilegia vulgaris*). The spurs on this plant are nectar-rich.*

*Above: Globe thistle (*Echinops ritro*) is very attractive to bees and other insects.*

*Above: Lemon balm (*Melissa officinalis*). Bees are drawn by the smell and nectar.*

Above: Nigella damascena 'Persian Jewel' is very attractive to bees.

*Above: Astilbe (*Astilbe ardensii sanguinea*) is a colourful addition to a bee garden.*

*Above: Borage (*Borago officinalis*) has an abundance of nectar.*

*Above: Shrubby cinquefoil (*Potentilla fruticosa*) is a good nectar source.*

- Fennel (*Foeniculum vulgare*)
- Foxglove (*Digitalis purpurea*)
- Globe thistle (*Echinops ritro*)
- Honeysuckle (*Lonicera periclymeum caprifoliaceae*)
- Lemon balm (*Melissa officinalis*)
- Love-in-a-mist (*Nigella damascena*)

- Manuka (*Leptospermum scoparium*)
- Meadow thistle (*Cirsium dissectum*)
- Potentilla (*Potentilla tormentil*)
- Purple loostrife (*Lythrum salicaria*)
- Rosebay willowherb (*Epilobium angustifolium*)
- Snapdragon (*Antirrhinum majus*)

- Snowberry (*Symphoricarpos albus*)
- Sunflower (*Helianthus annuus*)
- Teasel (*Dispacus japinoca*)
- Thyme (*Thymus vulgaris*)
- Vipers bugloss or blue thistle (*Echium vulgare*)
- Wood bettany (*Stachys officinalis*)

*Above: Fennel (*Foeniculum vulgare*) is an aromatic herb with nectar-rich flowers.*

*Above: Rosebay willowherb (*Epilobium angustifolium*) attracts a bumblebee.*

*Above: The nectar-rich honeysuckle (*Lonicera periclymeum caprifoliaceae*).*

AUTUMN PLANTS

The autumn season can be another difficult time for bees as spring and summer flowerings decline, but there are still some excellent plants available to provide nectar and pollen.

You may notice that some of the plants in the list below have already been mentioned in the spring flowering section. This is because some plants flower twice in one year in some parts of the world, where the climate is beneficial. Good examples of plants with this repeat-flowering trait are buckwheat and eucalyptus. Gorse is particularly useful because it can flower again for a good autumn and winter supply of pollen.

If you can take your bees to an area where there is heather, this is an exciting time. The autumn heather crop is a boon to many beekeepers and all part of their management plan. Sadly, like so

Above: The giant flower heads of Angelica (Angelica archangelica) are rich in nectar.

Above: Michaelmas daisy (Aster novi-belgii) provides a good nectar supply.

many plants, it depends on the weather, and can fail to produce nectar if the prevailing climatic conditions are not right.

Below are some more useful late summer and autumn plants that will also provide welcome colour in your garden, as well as attracting honey bees. The best plants for pollen and nectar are gorse and mahonia.

- Angelica (*Angelica archangelica*)
- Aster (*Aster* spp.)
- Buckwheat ((*Eriogonum douglasii*))
- Butterfly bush (Buddleia davidii)
- Cardoon (*Cynara cardunculus*)
- Cornflower (*Centaurea*)
- Dahlia (Dahlia pinnata)
- Delphinium (*Delphinium* spp.)

Left: The blue flowers of cornflower (Centaurea cyanus) are attractive to bees.

Above: Eryngium giganteum *has a long flowering period and thus a lot of nectar.*

- Fuchsia (*Fuchsia* spp.)
- Globe thistle (*Echinops ritro*)
- Golden rod (Solidago virgaurea)
- Gorse (*Ulex europaeus*)
- Heather (*Erica cinerea*)
- Holly (*Ilex aquifolium*)
- Hyssop (*Hysoppus officinalis*)

Above: Tingiringi gum (Eucalyptus glaucescens).

- Ivy (*Hedera helix*)
- Late clover (*Trifolium repens*)
- Lavender (*Lavandula officinalis*)
- Cantaloupe melon (*Cucumis melo*)
- Oregon grape (*Mahonia aquifolium*)
- Penstemon (*Penstemon barbatus*)
- Pumpkin (*Curcurbita pepo*)

Above: Fuchsia. The single flowers are sought by foragers for pollen and nectar.

- Scabious (*Scabiosa lucida*)
- Sea holly (*Eryngium giganteum*)
- Smooth Sow-Thistle (*Sonchus oleraceus*)
- Stonecrop (*Sedum*)
- Tingiringi gum (*Eucalyptus glaucescens*)
- Tall verbena (*Verbena bonariensis*)

*Above: Common holly (*Ilex aquifolium*)*

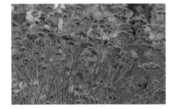

*Above: Tall verbena (*Verbena bonariensis*) flowers longer than most nectar plants.*

*Above: Penstemon (*Penstemon barbatus*) is very popular with bees and bumblebees.*

*Above: Stonecrop (*Sedum herbstfreude*) is very attractive to pollinating insects.*

PLANTS AND HONEY FLAVOURS

Honey bees feed on myriad nectar-bearing plants, and this makes quite a few interesting and delicious variations in the flavour of the honey that they produce.

Plants grown on different soils, in different climates and at different times of the year will produce honey of markedly varying flavours. Try a taste test with some clover honey followed by manuka honey, or tupelo honey followed by almond honey, and you will certainly notice an amazing difference.

Unfortunately so many people are used to buying sanitized, ultra-filtered, overheated, crystal clear honey at the supermarket that they do not appreciate the flavour of fresh honey any more. Supermarket honey resembles sugar syrup and has none of the individuality of local honey from an artisan beekeeper. But once people are aware of the exciting flavours and beneficial effects of unadulterated honey, they can quickly be converted.

Some common honey tastes are listed opposite. Try them and see what you think. If you can, purchase honey direct from a beekeeper rather than from a shop.

TRY A TASTING SESSION

In a recent experiment in New Zealand, a beekeeper gave away natural 'bush honey', filtered through a normal kitchen sieve, to his work colleagues and neighbours, many of whom had said they did not like honey. All were amazed that they could actually taste the flowers, and all of them found it delicious. After your harvest, why not have your neighbours round for a sampling session and demonstrate to them the real taste of honey.

Above: Three shades of clear honey.

Above: Blossom honey

Above: Lavender honey

Above: Greek pine honeydew

HONEY	CHARACTERISTICS
Wildflower	Varies according to the wildflowers in bloom, but you can definitely taste the flowers in the honey.
Clover	A popular honey with a mild, pleasant, sweet but non-sugary taste.
Buckwheat	Strong, dark in colour, and pungent, with a definite aftertaste.
Locust	A light, good quality honey with a distinctive, strong aromatic flavour.
Orange	A light honey with a delicate orange blossom flavour.
Thyme	Another aromatic honey with fairly robust herbal flavour with hints of thyme.
Eucalyptus	Varies in colour and flavour but is overall quite strong, with a medicinal aftertaste.
Manuka	Dark and rich, with a definite medicinal taste. This honey used to be either fed back to the bees or discarded until its extraordinary medicinal properties were discovered.
Honeydew	Taken from other insects such as aphids, which suck up plant juices, this honey is very dark and full of flavour. It is not popular in the US, where it is classed as 'baker's honey', but loved in Europe.

TAKING BEEKEEPING FURTHER

Beekeeping can be many things to many people, and is not just about producing honey. Most beekeepers will start in this way as a hobby, but having produced their own delicious honey for a couple of years, they often begin to venture into other beekeeping activities, such as entering association shows and displaying or selling their products. The educational side is important too. When beekeepers investigate further opportunities, such as harvesting other hive products or breeding queen bees for sale, they will find that they can continue to learn more about their fascinating charges.

Left: This garden has been planted with perennial plants to attract bees and other pollinating insects throughout the year.

BEEKEEPING INDUSTRIES

Once you start beekeeping, you will find that it is a fascinating hobby that you can go on exploring for life, or even turn into a profitable business.

Honey is just one of the products of the hive. Pollen, beeswax, propolis, royal jelly and even venom can all be harvested.

Hive products

Everything that bees produce has a ready market. The various harvests need different methods of hive and apiary management to maximize production.

You may also wish to investigate the world of queen production for your own hives, or to sell to other beekeepers. Breeding bees that have special characteristics can be very profitable. Even if you don't intend to sell your queen bees, breeding can be one of the most interesting aspects of beekeeping and can take you into the realms of genetic studies and the production of the 'perfect bee'.

Above: A bee is shown with one of the products of the hive, propolis. It is a red, sticky, odorous paste that is used to seal any gap in the hive.

Above: The harvest of the hive is not just honey; wax, comb, royal jelly, propolis and pollen can all be exploited in some way.

Another profitable activity is hiring out your bees to pollinate crops and taking the hives to the orchard or farm. The advantage is that the beekeeper will get honey from specific types of blossom, which can then be sold.

Associated activities

There are of course other industries that are associated with beekeeping. You can use your hive products to make cosmetics, beer and mead, as well as a host of other products. Some beekeepers venture into the world of apitherapy, which is a huge subject in itself. Salves, ointments and medicines using honey as a base have a very long history, and these products are beginning to enter mainstream medicine. Manuka honey

Above: Golden grains of pollen spill out of a jar. The pollen is mixed with bee saliva and nectar and formed into these pellets. Pollen is used by humans as a food supplement.

Above: Propolis capsules are reputed to have beneficial health effects.

and buckwheat honey have clinically proven medicinal properties, which have been shown to be superior in many ways to their chemical equivalents. Some of the honey treatments available are now beginning to be used in clinical medicine all over the world.

Beekeeping products

The industries associated with beekeeping, such as beehives, specialist tools and clothing manufacture, abound in all countries that have beekeepers.

Without these local specialist suppliers, beekeeping would become a difficult and more costly process.

The professional beekeeper

Beekeeping can be taken a stage further by going commercial. This is a rather different type of beekeeping, in which the enjoyment of a hobby is sacrificed to produce an income. The magic, beauty and mystery of the queen bee can be lost – she becomes merely the chief production unit of the colony,

which must be replaced as often as required to keep up with the beekeeper's demand for honey.

Sitting in a deck chair next to two or three hives on a balmy summer's evening, watching the bees come and go and relaxing with a glass of wine or a cup of tea, is exchanged for a hectic round of hard labour as you struggle to complete apiary inspections before nightfall and collapse exhausted into bed. On the other hand, it can be quite lucrative, and you will keep fit and sleep well.

Above: Beeswax is another by-product of the hive, which is prized for making smoke-free church candles, beauty products and hand creams.

THE SCIENCE OF BEEKEEPING

Like most farming and livestock industries, beekeeping needs scientists and researchers. Only by combining current scientific discoveries with the practical knowledge of beekeepers can this industry make advances in the future. Bees are vital to the wellbeing of the planet, and scientists now recognize that.

Above: A honeybee deposits nectar in drawn out comb.

POLLEN

Beekeepers have always been interested in this amazing substance, especially in the way it is produced by flowering plants, why bees collect it and how they use it within the hive.

Pollen is something of a super food, containing vitamins, protein and minerals. It benefits bees and humans.

Characteristics of pollen

Pollen is produced by the male flower's anthers in the form of a fine powder, and contains the microgametophytes of seed plants, which produce the male gametes (sperm cells). When the anthers of the male flower open, the pollen is transported by various means to the stigma of a receptive plant, and so pollination occurs. During this phase, the microgametophytes are protected by a strong pollen case, which can

withstand the effects of weather and age for many years. This movement of pollen grains can be accomplished by the wind or by insects, depending on the plant. Certain insects such as bees will be attracted to the flower by its nectar. Pollen grains will then adhere to the hairy body of the bee, and when it moves to a receptive flower to obtain more nectar, the transfer will take place.

Bees as pollinators

Bees are designed to carry pollen around, whether accidentally or by their own design. It is not simplifying the case too much to state that

pollination is really the intended purpose of bees in the natural order. They are the only insects that can collect pollen, and are unique in the insect world in that they feed pollen to their young. It is nutritious, containing all the food groups the bees need.

Bees evolved at the same time as flowering plants, and in doing so they have developed into specialized pollen carriers. Unlike wasps, for example, bees have plumose hairs to help them carry pollen. Their bodies also have a slight positive charge, while pollen grains have a slight negative charge, which attracts the pollen to the bee's body. Bees tend

Above: This bee is heavily laden with pollen. However, pollen is fairly light, and the worker bee will flit from flower to flower, gathering nectar and in so doing, she will collect pollen, too. The bee is the only pollinator that actually harvests pollen.

Above: The pollen sacs or baskets are clearly seen on this bee's legs.

Above: A few pollen grains remain at the entrance to the hive.

Above: Pollen grains are spread out to dry before packaging.

to remain constant to one type of flower as long as the nectar source is good, so they can pollinate an area of the same plants very rapidly and efficiently. For this reason, bees are essential as pollinators for our agricultural crops, especially since the habitats of other insect pollinators are often destroyed by that very same agricultural effort.

Pollen as a hive product

The small pellets of pollen, which you will see many bees carrying into the hive as protein food for the brood, are carefully collected by the bee from many flowers, 'combed' from its body hairs and packed (with the addition of a tiny amount of nectar) into the corbicula, or pollen basket, on its hind legs. Bees have developed sturdier hind legs than other insects to carry this extra weight. These small pellets are unloaded by the bee into cells around the honey cells of the brood chamber, where they will be used for brood food.

Collection and use of pollen

In commercial enterprises, beekeepers can collect the pollen grains by inserting a pollen trap in the beehive. This simple device is a small grid placed over the

hive entrance, through which the bees clamber on entering. The grid scrapes the pollen grains off the bees' legs and they drop into a collection container under the hive or in front of the entrance. Once it has been collected, the pollen is carefully dried until the

moisture content is at the correct low level, and then it is packed and sealed, ready for sale to a very receptive market as a natural health food.

Remember that the bees need their pollen too, so do not leave a pollen trap on one hive for too long.

Above: Pollen grains are caught in a pollen trap that the beekeeper has installed to scrape the pollen from the baskets on the bees' legs as they enter the hive.

PROPOLIS

This substance is a resinous mixture that bees collect from tree buds, sap flows or other botanical sources, and is one of the earliest-known medicines used by man.

Owing to its amazing antimicrobial qualities, propolis has been claimed to be beneficial to humans as well as bees.

Characteristics of propolis

The composition of propolis varies tremendously, and depends entirely upon the source. Propolis from one source may also change its composition with the seasons. Usually it is dark brown in colour, but it can be found in green, yellowish brown, red, black and white hues, depending on the sources of the resin. At room temperature and above it remains sticky, but below this it starts to harden and become brittle.

Bees use propolis to glue parts of the hive together and to seal any gaps. Because it is an antiseptic, it also maintains hygiene in the hive.

How bees use propolis

There are many ways in which bees use propolis: to reinforce the structural stability of the hive, to make it more defensible by sealing alternative entrances, to prevent diseases and parasites from entering the hive and inhibit viral and bacterial growth, and to prevent putrefaction within the hive. They can also use propolis to mummify hive intruders such as mice that may die in the hive and are too big for the bees to carry out.

The bees gather propolis by chewing the sticky resin from the bark of trees, leaf buds and flowers, until it reaches the correct consistency to take back to the hive. They transport it to the hive in the same way that they carry pollen, by attaching it to their corbiculae or

pollen baskets and carrying it back to the hive. Hive bees help them to unload the propolis in the hive. The bees use the propolis soon after collecting it, and will start gluing the hive boxes together, which is why you need a hive tool to separate them for inspection. They also tend to glue the frames together, as well as any other gaps. Some bees tend to use more propolis than others; a lot can cause problems for the beekeeper unless it is going to be harvested.

Commercial uses of propolis

Many beekeepers scrape propolis off the frames and boxes and throw it away, but it does have value. It is used in paints and varnishes (it was reputedly used by Stradivarius on violins), as well as

Above: The bees use propolis as an antiseptic on the walls and flat surfaces in the hive, to keep infection at bay.

Above: These chunks of propolis have been scraped from various areas in the hive where the bees used this sticky resin.

Above: A hive tool is invaluable when you need to scrape propolis from the joints and frames in the hive.

Above: Bees will cover netting on the inside of the roof with propolis.

Above: This type of propolis screen is made from plastic.

Above: A propolis screen placed under the crown board will also collect propolis.

medicinally in dentistry, as an anti-bacterial, anti-fungal and anti-viral topical liquid and spray. It can even be used in chewing gum.

Collection and storage

There are two ways of collecting propolis. The first method is simply to scrape the propolis off the gaps between frames and other surfaces with a hive tool or other device, and place it in a container. Try not to accidentally scrape off pieces of wood or other substances as well. Place the container in a cold area or refrigerator, and once the propolis is hard, separate out any foreign particles. The second (and easier) method is to place a 'propolis mat' over the top of the frames just below the lid or crown board. A propolis mat is simply a plastic or fabric mesh. The bees will fill the holes in the mesh with propolis. Once it is full, remove the mesh and place it in the refrigerator to chill. When the propolis has completely hardened, scrunch up the mesh over a container, and the pure propolis will fall out into the receptacle.

Propolis usually contains a lot of wax or wax particles, and stored propolis will be subject to wax moth predation. Keep it well sealed in a container to protect it from infestation.

Once you have got propolis on your hands or clothes, it is very difficult to remove.

Above: Scraping off the propolis is a good way to clean the hive boxes.

Above: To prevent it building up, it is a good idea to remove the propolis regularly, especially if you intend to use it commercially.

ROYAL JELLY

What exactly is royal jelly, and where did it get its magical reputation as a compound of life-prolonging elixirs, full of known and unknown vitamins and minerals?

The common misconception that royal jelly is a cure-all and will make an astounding anti-ageing cream or supplement is perpetuated by companies selling goods containing microscopic amounts of royal jelly, and claiming amazing results.

Food for a queen

Royal jelly is indeed an extremely rich food source for bees. It comes from the hypopharyngeal gland in the head of a worker nurse bee. It is acidic and rich in protein, vitamins, RNA, DNA, fatty acids and sugars. When you look at a tiny larva in its cell in the comb you will see that it appears to be floating in a milky white liquid. That liquid is royal jelly, and

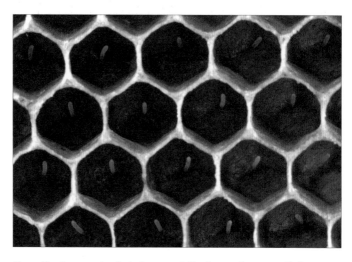

Above: After these eggs hatch, the larvae are fed by the nurse bees on royal jelly only for the first few days.

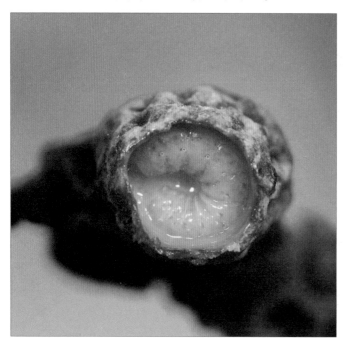

the nurse bee has deposited it there for the larva to absorb as food. Every bee larva is fed on this for the first three days of its existence, whether it is a queen, a drone or a worker.

A queen or not?

All female (non-drone) bee larvae are capable of becoming queen bees and are genetically the same. After three days of feeding royal jelly to all the larvae, the nurse bees limit the amount of royal jelly fed to the majority and replace part of the feed with honey and pollen. Royal jelly continues to be fed to those few female larvae destined to be queens. The egg-producing ovarioles will

Left: Any larva destined to be a queen continues to be fed with royal jelly after day three of its life.

Above: There is much commercial potential in the production of royal jelly. The unusual properties of this high-energy food has provoked a demand for supplements of royal jelly.

develop and the bee will eventually emerge as a queen. For the rest, the sharp reduction in royal jelly in the diet will prevent ovariole development and these will become the worker bees.

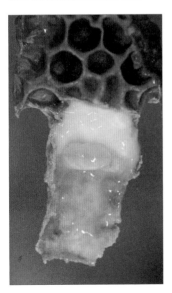

Above: A section through a queen cell reveals the larva bathed in royal jelly.

A simple dietary change that can effect such a dramatic alteration in form and function between genetically identical larvae does indeed seem magical, and is an impressive feat of nature. The scientific term for this is 'environmentally controlled phenotypic polymorphism'.

Magic for humans

It is this aspect of major change caused by diet that has so impressed observers over the years. What could a diet of royal jelly do for humans, if it can produce a queen bee? Because of this, weight for weight, royal jelly is probably the most valuable hive product, and many commercial beekeepers focus solely on the production of royal jelly.

Apart from its debatable anti-ageing properties, there is preliminary evidence through medical research that royal jelly may have some worthwhile effects in lowering cholesterol, healing

inflammatory wounds and fighting bacterial infections, as well as protecting the immune system and producing increased energy.

Harvesting royal jelly

A strong hive, bulging with bees, is needed to produce many queen bees. Once the nurse bees have deposited a large amount of royal jelly in each queen larva cell, the beekeeper removes the larvae and sucks out the royal jelly from each cell using a tiny suction device. China produces most of the world's royal jelly, and there, the jelly is often scooped out by hand using tiny wooden spoons.

Once harvested, the jelly is either frozen or freeze-dried and packed ready for sale. A large number of hives are required to harvest worthwhile amounts of royal jelly, so its production is usually restricted to the commercial sector.

BEESWAX

From ancient times there have been many uses of beeswax, and it is still used commercially to make fine candles, cosmetics and pharmaceuticals, as well as polishes.

Worker bees of a certain age synthesize sugars obtained in honey into beeswax, which they then extrude through glands underneath their abdomen. This is a colourless liquid which hardens into a small clear plate or scale of wax on its 'wax mirrors'. Each worker has four of these mirrors underneath its abdomen. The wax scales are about 3mm ($\frac{1}{10}$in) across and 0.1mm ($\frac{4}{1000}$in) thick,

and about 1,100 of the scales are required to make just 1g ($\frac{1}{25}$oz) of beeswax.

When you are inspecting your bees, you will at times find many of them clinging together in chains between the frames. These bees are producing wax in the warm environment of the hive. The ambient temperature in the hive has to be 33–36°C (91–97°F) for this activity.

Uses of beeswax

Nowadays, cosmetics and pharmaceuticals account for 60 per cent of the total industrial production of beeswax. It is also used in many polishing materials and is a component of modelling waxes.

Beeswax as a hive product

Beekeepers should always ensure that they replace old, black comb on a three-year cycle, and allow the bees to draw out new comb to replace it. By rendering down old comb, beekeepers maintain the cleanliness of the hive and produce wax for sale or re-use.

Bee supply companies may buy surplus beeswax, and many will swap it for foundation wax – a cost-effective way of obtaining new foundation. Beekeepers who use plastic frames can dip them in molten wax to encourage bees to draw out, or make their own candles quite simply with the aid of a good instruction book.

Rendering wax

There are several ways to render beeswax, but for the hobbyist the best way is to make a solar wax extractor. This is a simple device in the form of a large box with a glass cover. Place old wax, whether in the frame or not, in a tray in the extractor and leave it out in the open with the box tilted towards the sun. The old comb melts and drains through a grid into a flexible plastic container in the box. Once this container is full it is removed, the wax hardens and you can turn it out as a

Above: Sheets of beeswax can be rolled into candles around a wick. Beeswax candles burn longer and brighter and they do not smoke in the way that paraffin candles do.

Above: Beeswax is used to make fragrant polishes for wood and leather.

block of clean wax. The old comb left behind, known as slumgum, consists of larval cocoons which still contain wax. This remaining wax can be extracted in a steam extractor, but for many hobby beekeepers this equipment is rather expensive, so they use the remains of the old comb as very effective, and smokeless, firelighters.

Above: A solar wax extractor is a glass covered insulated box, which is used to melt the old wax from the combs and cappings. It is then drained into plastic containers.

Above: Different shades of beeswax, from light to dark, can be produced.

Above: Beeswax, with the addition of oatmeal, has been used to make a soap.

Above: The old wax cappings can be collected and reused to make beeswax.

VENOM

Venom is a very complex substance composed of various peptides, amines and enzymes, which has evolved over millions of years as a defence for bees and bee colonies.

Bee venom is probably the main reason why people are afraid of keeping bees. Bees will use their stings if their nest is threatened (possibly another reason why swarms tend not to sting – they have no nest or brood to defend), or if they are panicked.

What causes the pain and swelling?

Venom can cause pain, venom poisoning and even death in allergic individuals. The venom is a colourless liquid that can be harvested and dried to become a white powder-like material. The main component of bee venom responsible for pain in vertebrates is the toxin melittin. This comprises about 50 per cent of the dry weight of venom, and it is assisted by another component, hyaluronidase, which breaks down cell membranes and helps the venom to spread. Histamine and certain amines may also contribute to pain and itching.

Above: The bee's sting, containing venom, can be clearly seen with a barb on the end. The barb can not be removed from the victim, hence the bee dies after stinging.

Above: When a bee stings near delicate tissues, it causes a lot of swelling.

Venom poisoning can occur if a large number of stings are received (usually 500–1,500). On rare occasions a single sting can induce anaphylactic shock in a victim, which can cause death within minutes.

It is a wise move to explain to your doctor that you are a beekeeper and obtain a prescription for an EpiPen, which is a small auto-injector of epinephrine that can save your life if you go into anaphylactic shock. Anyone can become sensitized to bee venom, and even though it is rare, anaphylactic shock can kill. Beekeeping companies in many countries are obliged by health and safety regulations to provide EpiPens for their beekeepers in the field.

Uses of venom

Bee venom is a source of pharmaceutically active components and is used in many products in the drug industry, especially in Europe. These products include creams, liniments, ointments, salves or injection forms for treating different human ailments. Recently, scientists researching a toxin extracted from the venom of the honey bee have begun to develop new treatments to alleviate the symptoms of conditions such as muscular dystrophy, depression and dementia.

Harvesting venom

Venom can be harvested by beekeepers by using an electrical apparatus which induces bees to sting when they come

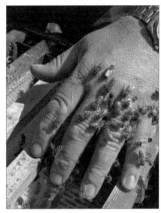

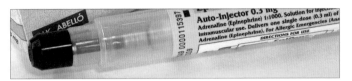

Above: This pen is one of many brands that are used in the event of anaphylactic shock from several bee stings.

Above: An EpiPen, containing adrenaline (epinephrine), is used as a single dose in an emergency such as anaphylactic shock.

Above: This beekeeper supplies bees for venom therapy.

into contact with it. A collector frame is placed at the entrance of the hive and connected to an electrical supply. The frame is made from wood or plastic, and holds a wire grid carrying the current. Underneath the wire grid is a glass sheet which can be covered with a plastic or thin rubber membrane to avoid contamination of the venom.

When they land on the alighting board of the hive, the bees come into contact with the wire grid and receive a mild electric shock, whereupon they sting the surface of the collector sheet. The venom is deposited between the glass and the protective material, where it dries and is later scraped off. The membrane through which the bee stings is too thin to detain the bee stinger, so the bee is able to remove it and live. Like the producers of royal jelly, successful bee venom producers tend to be commercial operators with many beehives.

TAKE ACTION WHEN STUNG

If a bee stings you, the venom is accompanied by an alarm pheromone that will attract other bees to head for the same spot and sting again. Move away quickly. Remove the sting by scraping with a fingernail, wash the site with soap and water, and place ice on the sting to reduce swelling and relieve the pain.

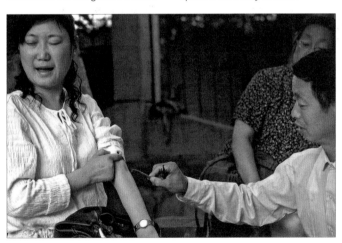

Above: A woman in China receives treatment with bee venom for rheumatism. The venom is thought to cure diseases such as psoriasis, arthritis and high blood pressure.

Above: Bees are bred to inject patients with venom, after which the bee dies.

SHOWS AND COMPETITIONS

Shows and competitions are one of the main highlights of the beekeeping year for many beekeepers – they can be a social gathering and competition all in one.

Entering your honey at a show is an ideal way of judging how good your product really is, and is a place where you may be able to pick up tips and advice from experienced beekeepers.

The annual show

If you join your local beekeeping association, you will undoubtedly be persuaded to take part in the annual show and competition. Many novice beekeepers are reluctant to do so at first because they think that their entry may not meet the exacting standards required, but of course the best way to learn about these standards is to enter the competition and learn from your mistakes, as well as to compare your product with others. Another benefit is that you will meet other beekeepers and can compare notes with them.

One thing you must remember is that when you decide to start entering these competitions, you are entering a world of experienced beekeepers, so the competition is keen. If the level of honey in your glass is a tiny fraction of a fraction below the 'standard', for example, you will be disqualified. So remember to be determined and do not be discouraged if it takes a while to win a prize.

Local shows

These can be great fun, and you will learn a lot about how to display your products in the best possible light.

From the local show, you can move up to trying for an entry in a national honey show such as the annual London show, or even Apimondia, the conference of the World Federation of

Beekeepers' Associations, which is held every two years, each time in a different capital city around the world. This event attracts thousands of beekeepers, scientists and business people, as well as the general public.

Classes and entry requirements

You will find that there is a competition class for just about everything, depending on how large the show is. You might win a prize in any one or more of these categories: wax blocks of all shapes and sizes; light honeys; dark honeys; set honeys; creamed honeys; liquid honeys; books; magazines; beekeeping inventions; honey cakes and other foods containing honey; meads and honey beers. Some people make decorations using beeswax, and of course wax candles.

Above: Prize-winning jars of honey are on display at a smallholders annual show. There are two shades of clear honey and some creamed honey.

Above: Different shades of mead, or honey wine is displayed at a honey show.

LEARNING FROM THE EXPERTS

Local beekeeping associations usually give tips that will help you to excel in competitions. For example:
• Honey for show purposes must be finely strained to avoid the presence of dust or other fine particles.
• If extracting with a hand-operated machine, operate it at a low speed to avoid a build-up of fine air bubbles, which are impossible to eliminate later.
• Show honey should be extracted only from virgin comb to obtain a good sparkle.

Above: A little girl samples some sweets at a honey show, just one of the many foods made from honey that are on display.

Above: Other products of the hive, such as items make from beeswax, are also displayed in honey shows, such as these rolled wax candles.

The entry requirements and costs of these competitions and shows are spelled out in great detail well in advance, and to stand a chance of winning a prize, you must adhere strictly to all these requirements. The guidelines include using scrupulously clean jars to bottle your honey, and the honey must not contain any foreign bodies. Any impurities will be revealed by the judge's torch.

It is also vital to get your entry in early. This may all sound quite daunting to the beginner, but if you start off at local association level you will have a lot of fun, as well as meeting some interesting people, who will probably be happy to give you some good, free advice, and you will be inspired by the winners.

Above: Different shades of wax are on display at a honey show.

Above: Here, beeswax has been poured into moulds and set.

QUEEN REARING

As a way of increasing your enjoyment of beekeeping and learning more about the biology and development of bees, queen rearing is very much recommended.

Why raise queens?

Queen bees can be purchased from any reputable breeder, and there are sufficient of these to meet market demand, but there are also good reasons for raising your own. There may be some superior characteristics in your stock, and you want to ensure that these are perpetuated; you want to learn more about the biology of bees; you want to be able to sell queens – many beekeepers like to have a local supplier.

Above all, it is cheaper than buying in queens, and it is a simple process, not a mysterious and complicated business as some novice beekeepers think.

How a queen is produced

Rearing queen bees is based entirely upon the observation of one fundamental principle of bee biology. Nurse bees can transform a larva that is up to three days old into a queen by enlarging the cell and feeding it a plentiful supply of royal jelly until the cell is capped. So, if you are able to introduce these tiny larvae, already in enlarged cells, to a colony which has no queen, and thus has many highly motivated nurse bees, they will raise queens.

Although it is an easy process, queen rearing can involve some delicate larval grafting. An easier way for the novice beekeeper is to use the a queen rearing kit such as the Jenter system (or similar). These queen-rearing kits are well worth buying, and basically consist of a plastic frame of embossed cells, a queen excluder and, at the rear, small cell plugs. The kits come with full

Above: Worker bees in the process of constructing a queen cup.

instructions, and it is this system that is featured here. Queen-rearing systems depend on your preparing nucleus hives at various stages. Each nucleus hive should contain two frames of brood (sealed); two frames of stores (honey) and many bees. Make these up from the

Above: Honey bees are seen on queen-rearing cells.

hive that you have chosen to rear queens from. (See Step 1.)

Prepare a nucleus hive during the evening prior to day four (Steps 4, 5 and 6.) Then prepare your nucleus boxes on day 8 to take the queen cells on day 13 (Step 8.)

QUEEN REARING

1 Choose a populous hive with a gentle, healthy queen. From the centre of the brood box, remove a frame of brood. Make sure the queen is not on it. Shake off the bees into the hive. Using pliers, cut out a section as shown.

2 Insert the Jenter box into this cut-out section of the frame, and attach it securely with wires. The bees will soon fix the cage firmly into the comb with wax. The cage is now ready for the insertion of the queen.

3 *Insert the frame (now including the Jenter box) into the brood chamber. Replace it in the same position it was in before and leave it for the bees to draw out comb. (Some kits simply plug on to the brood frame without the need to cut out a section). Catch the queen using a queen catcher, available from bee supply stores, and place her into the box. Keep her there for a couple of days while she lays eggs in the cells. She can then be released back into the hive.*

4 *On day four, the eggs will have hatched and the tiny, comma-shaped larvae at twelve hours old will be seen floating in beds of royal jelly in the base of their cells. Nurse bees will be attending to these cells, ensuring that the larvae are well fed and always have sufficient food. Always handle the frame with extreme care at this stage, and when inspecting it, avoid exposure to strong sunlight, wind or cold temperatures.*

5 *Now remove however many plastic cells you need (if you need 5 queens, remove 7 or 8 cells, as some may not be successful) from the rear of the cage and, using the plastic plugs that come with the kit, attach them to a prepared cell bar on an otherwise empty frame. This is not a complicated procedure and full instructions are provided with each kit. Treat each removed cell with great care when attaching them to the cell bars.*

6 *Place this cell frame into a prepared nucleus containing four other frames – two of stores (honey) and two of well sealed brood (no open brood). Shake into this nucleus many young nurse bees from the brood frames of the hive. It is important to ensure that there are a lot of nurse bees; they are vital and they are always to be found on brood frames in the main colony.*

7 *Nine days later, gently remove the frame from the nucleus box. You should now have a series of well-developed queen cells ready to be transferred to your nucleus hives. Not all of the cells will have been capped over, and these should be discarded as failures. It happens, even to experienced queen rearers. This is the reason more cells than are required should be prepared initially.*

8 *Hold up the frame and you should see many nurse bees surrounding your new queen cells. Very gently brush off the bees and, equally gently, unplug each cell from the bar. You can now go to your prepared nucleus hives and press each queen cell into the centre of a frame of brood. Ensure that the cell is facing downwards. Place the frame back into the nucleus and close up.*

QUEEN BREEDING

Encouraging certain positive traits in queen bees also means breeding these traits into the entire colony, and this is the main reason that beekeepers breed their own queens.

Breeding any form of livestock means more than mere rearing. Rearing is the procedure of producing queen honey bees, whereas breeding is the process of selecting certain advantageous traits and breeding them into the next generation and beyond.

Breeding desirable characteristics

In honey bees there are many traits that you may wish to develop in this way, including: gentleness; high egg-laying rate; low-swarming tendency; hygienic behaviour (which assists hugely in the avoidance of disease); an ability to suppress *Varroa* mite reproduction; the quality of being able to forage in bad weather conditions; and many more. Other considerations may apply to your queen-breeding. You may decide, for example, that you want to breed bright yellow bees or jet black ones. You might even be trying to breed the 'perfect' bee – the one that doesn't sting, lays millions of eggs, produces tons of honey,

never swarms and is completely disease-resistant. You can do this if you have patience and the ability to select the right breeding stock. Government laboratories all over the world are engaged in just these activities with honey bees, and in these days of dwindling bee populations their work is becoming ever more important.

Queen breeding can be done by any beekeeper who has learned how to rear queens. The same fundamental principle applies: if sufficiently motivated by the lack of a queen, nurse bees will turn three-day-old larvae into queens. If you have successfully reared queens following the eight-step system on the previous page, you can breed them too.

Selecting characteristics

The previous page has already introduced a simple procedure for breeding queen bees. You select the parent colony and use your favourite

Above: Using a spatula, a beekeeper recovers honey bee eggs from cells in the hive for breeding queen bees.

queen to lay the eggs in the frame that was introduced to the nurse bees. You no doubt selected this queen because the colony produced a good honey crop, or because the bees were

Above: The honey bees shown here are bred for resistance to the bee mite.

Above: The beekeeper grafts bees to breed a mite-tolerant specimen.

Above: A honey bee breeder picks out a suitable bee larva for breeding.

gentle, or perhaps both. You may in the future wish to select a queen based on a few other advantageous characteristics. Follow these simple rules for selecting the queen for breeding:

• Decide which characteristics you would like to propagate.

• Decide how important each characteristic is to you. For example, if you live in an urban area and gentleness is twice as important as honey productivity, this trait would score twice as highly as an average yield of honey.

• Give each of your colonies a score over the year and add them up.

• Choose your breeder queen from the colony that scored the most points.

Queen breeding depends on having a large selection of hives from which to choose the best breeder queens. The average hobbyist will have more

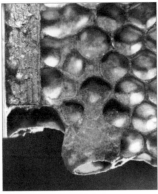

Above: A queen cup is a cup-like precursor to a queen cell. The cup is empty but the cell contains an egg.

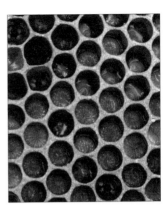

Above: When the eggs have hatched, the tiny larvae can be transferred to your prepared queen cups.

limited possibilities. However, many associations have groups of beekeepers who co-operate with each other in the business of queen breeding, and so are able to develop excellent strains of

bees for their local areas. It is well worth approaching your local association to see if such a group exists; if it does not, you might be able to interest enough beekeepers to set one up yourself.

Above: A low-swarming tendency would be an ideal trait to breed into a queen. If the colony is calm and settled, it will have a higher productivity.

Above: A beekeeper who has tried to breed some queen bees examines a frame to check whether there are any queen cells.

SCIENCE AND BEE RESEARCH

Every country in the world that has honey bees has at least one laboratory dedicated to researching them, and public interest in supporting this research is at an all-time high.

Honey bees are the world's most studied insect. This is because they are also economically the world's most important insect, and while their usefulness is huge, their numbers are dwindling. It is difficult to overestimate the importance of the honey bee both to the global economy and to the global food supply. These factors add up to billions of pounds, euros and dollars being spent on research.

Studying bees

Because of this high interest and government dependence on the agricultural sector all around the world, there is always a place for new researchers in many of the laboratories. It is yet another aspect of beekeeping where young men and women can make a career if they are interested and qualified in the relevant subjects.

Science and the future

In the last few years, an international consortium of researchers announced that it has finished sequencing the entire genome – all the DNA – of the honey bee. This has opened many new avenues of exploration for scientists.

One example is the use of bee venom. Earlier in this book, medical research into the substance was mentioned. By using very high-technology research methods and computer models, alongside a genetic approach, research into alleviating several human diseases with the use of bee venom has been accelerated.

Another excellent example of very recent research which can both benefit the human condition and help bee breeders to produce better honey bees is outlined below.

An investigation compared the antibacterial activity of medical-grade honey in test tubes against a panel of antibiotic-resistant, disease-causing bacteria. The researchers developed a method to selectively neutralize the known antibacterial factors in honey (such as hydrogen peroxide) and

Above: Mating hives are shown outside the laboratory of Sussex University Honeybee Laboratory, UK. Research takes place here to study all aspects of the honey bee.

DON'T BE PUT OFF

One bee scientist in the UK studied bees, and lectured to appreciative audiences in colleges and schools, yet he was fatally allergic to bee venom, and so rarely, if ever, went near bees. Every interested scientist can become a researcher, even with the most unlikely handicap. Because there is a vital need to discover more about why bees are declining, more scientists are now involved in research.

Left: An apiarist checks one of the beehives at a research farm.

determine their individual antibacterial contributions. They then isolated what is called the defensin-1 protein, which is part of the honey bee immune system and is added by bees to honey. After analysis, the scientists concluded that the vast majority of honey's antibacterial properties come from this protein. This information also sheds light on the inner workings of honey bee immune systems, which may one day help breeders create healthier honey bees.

If you have an interest in developing research into bees, don't hesitate. There is a future for scientists and researchers into honey bees, their traits, their future health and the use of their hive products by humankind.

Above: The honey bee laboratory walls have coloured entrances to guide bees into hives that are inside.

Above: Australian Manuka and Jellybush honey has ULF (Unique Leptospermum Factor) activity that is equivalent to the Unique Manuka Factor rating given to New Zealand Manuka honey. The ULF rating is an indicator of the strength of the antibacterial effect.

GLOSSARY

Alighting board: a small projection or platform at the entrance of the hive.

American Foul Brood (AFB): a brood disease of honey bees caused by the spore-forming bacterium, *Bacillus larvae*.

Anaphylactic shock: constriction of the muscles surrounding the bronchial tubes of a human, caused by hypersensitivity to venom, and resulting in death unless immediate medical attention is received.

Apiary: colonies, hives, and other equipment assembled in one location for beekeeping operations; bee yard.

Apiculture: the science and art of raising honey bees.

Apis mellifera: the scientific name of the honey bee.

Bait hive: a hive placed to attract swarms.

Bee blower: an engine with blower used to dislodge bees from combs by creating a high-velocity, high-volume wind.

Bee brush: a brush or whisk broom used to remove bees from combs.

Bee escape: a device used to remove bees from supers and buildings by letting bees pass one way, but preventing their return.

Beehive: a box or receptacle with moveable frames, for housing a colony of bees.

Bee space: a tiny space between combs and hive parts in which bees build no comb or deposit only a little propolis.

Beeswax: a complex mixture of organic compounds secreted by special glands on the last four visible segments on the ventral side of the worker bee's abdomen, and used for building comb.

Bee venom: the poison secreted by special glands attached to a bee's stinger.

Bottom board: the floor of a beehive.

Brace comb: a bit of comb built between two combs to fasten them together, such as between a comb and adjacent wood.

Brood: bees not yet emerged from their cells: eggs, larvae, and pupae.

Brood chamber: the part of the hive in which the brood is reared; may include one or more hive bodies and the combs.

Capped (or sealed) brood: pupae in cells sealed with a porous wax cover.

Cappings: the thin wax covering of cells full of honey; the cell coverings after they are sliced from a honey-filled comb.

Castes: the three types of bees that comprise the adult population of a honey bee colony: workers, drones, and queen.

Cell: hexagonal compartment of honey comb.

Chilled brood: immature bees that have died from exposure to cold; commonly caused by mismanagement.

Cluster: a large group of bees hanging together, one upon another.

Colony: the aggregate of worker bees, drones, queen, and developing brood, living as a family unit in a hive or other dwelling.

Comb: a mass of six-sided cells made by honey bees, in which brood is reared and honey and pollen are stored.

Comb foundation: a commercially made structure consisting of thin sheets of beeswax with the cell bases of worker cells embossed on both sides

Creamed honey: honey which has been crystallized under controlled conditions.

Crystallization: *see* Granulation.

De-queen: to remove a queen from a colony.

Dividing: separating a colony.

Drawn combs: combs with cells built out by honey bees from a sheet of foundation.

Drifting of bees: the failure of bees to return to their own hive in an apiary containing many colonies.

Drone: the male honey bee.

Drone comb: comb used for drone rearing and honey storage.

Drone layer: an infertile or unmated laying queen.

Dwindling: the rapid dying off of old bees in the spring; sometimes called spring dwindling or disappearing disease.

Dysentery: an abnormal condition of adult bees characterized by severe diarrhoea, and usually caused by starvation or low-quality food.

European Foul Brood (EFB): an infectious brood disease of honey bees caused by *Melissococcus plutonius*.

Extracted honey: honey removed from

the comb by centrifugal force.

Fermentation: a chemical breakdown of honey, caused by sugar-tolerant yeast.

Fertile queen: a queen, inseminated instrumentally or mated with a drone, which can lay fertilized eggs.

Field bees: worker bees at least three weeks old that work in the field to collect nectar, pollen, water, and propolis.

Frame: four pieces of wood designed to hold honey comb, consisting of a top bar, a bottom bar, and two end bars.

Fructose: the predominant simple sugar found in honey; also known as levulose.

Fume board: a rectangular frame, the size of a super, covered with an absorbent material such as burlap, on which is placed a chemical repellent to drive the bees out of supers for honey removal.

Grafting: removing a worker larva from its cell and placing it in an artificial queen cup in order to have it reared into a queen.

Grafting tool: a probe used for transferring larvae when grafting queen cells.

Granulation: the formation of sugar (dextrose) crystals in honey.

Hive: a man-made home for bees.

Hive body: a wooden box which encloses the frames.

Hive stand: a structure to support a hive.

Hive tool: a metal device used to open hives, pry frames apart, and scrape wax and propolis from the hive parts.

Honey: a sweet viscid material produced by bees from the nectar of flowers.

Honeydew: a sweet liquid excreted by aphids, leaf hoppers, and some scale insects that is collected by bees.

Honey extractor: a machine which removes honey from the cells of comb by centrifugal force.

Honey flow: a time when nectar is plentiful and bees produce and store surplus honey.

Honey stomach: an organ in the abdomen of the honey bee used for carrying nectar, honey, or water.

Increase: to add to the number of colonies, usually by dividing those on hand.

Inner cover: a lightweight cover that is used under a telescoping cover on a beehive.

Invertase: an enzyme produced by bees to transform sucrose to dextrose and levulose.

Larva (plural, larvae): the second stage

of bee metamorphosis; a white, legless, grublike insect.

Laying worker: a worker that lays infertile eggs, producing only drones, usually in colonies that are queenless.

Levulose: *see* Fructose.

Mating flight: the flight taken by a virgin queen while she mates with several drones.

Mead: honey wine.

Melissococcus plutonius: bacterium that causes European Foul Brood.

Nectar: a sweet liquid secreted by the nectaries of plants; the raw material of honey.

Nectar guide: color marks on flowers believed to direct insects to nectar sources.

Nectaries: the organs of plants which secrete nectar, located within the flower (floral nectaries) or on other portions of the plant (extrafloral nectaries).

Nosema: a disease of the adult honey bee caused by the protozoan *Nosema apis*.

Nucleus (plural, nuclei): a small hive of bees, usually covering from two to five frames of comb and used primarily for starting new colonies, rearing or storing queens; also called a 'nuc.'

Nurse bees: young bees, three to ten days old, which feed and take care of developing brood.

Out-apiary: an apiary situated away from the home of the beekeeper.

Oxytetracycline: an antibiotic used to prevent American and European Foul Brood.

Package bees: a quantity of adult bees, with or without a queen, contained in a screened shipping cage.

Paralysis: a virus disease of adult bees which affects their ability to walk or fly.

Parthenogenesis: the development of drone bees from unfertilized eggs.

Play flight: short flight taken in front of or near the hive to acquaint young bees with their immediate surroundings.

Pollen: the male reproductive cell bodies produced by anthers of flowers, collected and used by honey bees as their source of protein.

Pollen basket: curved spines or hairs, located on the outer surface of the bee's hind legs and adapted for carrying pollen and propolis.

Pollen trap: a device for removing pollen loads from the pollen baskets of incoming bees.

Prime swarm: the first swarm to leave the parent colony, usually with the old queen.

Propolis: sap or resin collected from trees or plants by bees. Strengthens the comb, closes up cracks and keeps the hive healthy.

Pupa: the third stage in the development

of the honey bee, during which the organs of the larva are replaced by those that will be used by an adult.

Queen: a fully developed female bee, larger and longer than a worker bee.

Queen cage: a small cage in which a queen and three or four worker bees may be confined for shipping or introduction into a colony, or both.

Queen cage candy: substance made by kneading powdered sugar with invert sugar syrup until it forms a stiff dough; used as food in queen cages.

Queen cell: a special elongated cell, resembling a peanut shell, in which the queen is reared.

Queen clipping: removing a portion of one or both front wings of a queen to prevent her from flying.

Queen excluder: metal or plastic device with spaces that permit the passage of workers but restrict the movement of drones and queens to a specific part of the hive.

Queen substance: pheromone material secreted from glands in the queen bee and transmitted throughout the colony by workers to alert other workers of the queen's presence.

Rendering wax: the process of melting combs and cappings and removing refuse from the wax.

Revetment: a narrow ledge at the top of the hive body, from which the frames hang.

Robbing: stealing of nectar, or honey, by bees from other colonies.

Royal jelly: a glandular secretion of young bees, used to feed young brood.

Sacbrood: a brood disease of honey bees caused by a virus.

Scout bees: worker or forager bees that leave the hive to search for a new source of pollen, nectar, propolis, water, or a new home for a swarm of bees. They then direct the rest of the colony.

Self-spacing frames: frames constructed

so that they are a bee space apart when pushed together in a hive body.

Slumgum: the refuse from melted comb and cappings after the wax has been rendered or removed.

Smoker: a device in which hessian (burlap), wood shavings, or other materials are slowly burned to produce smoke which is used to subdue bees.

Solar wax extractor: a glass-covered insulated box used to melt wax from combs and cappings by the heat of the sun.

Spermatheca: a special organ of the queen in which the sperm of the drone is stored.

Sting: the modified ovipositor of a worker honey bee.

Sucrose: principal sugar found in nectar.

Super: any hive body used for the storage of surplus honey.

Surplus honey: honey removed from the hive which exceeds that needed by bees for their own use.

Swarm: the aggregate of worker bees, drones, and usually the old queen that leaves the parent colony to establish a new colony.

Swarming: the natural method of propagation of the honey bee colony.

Swarm cell: queen cells usually found on the bottom of the combs before they swarrn.

Uncapping knife: a knife used to shave or remove the cappings from combs of sealed honey prior to extraction; usually heated by steam or electricity.

Uniting: combining two or more colonies to form a larger colony.

Venom allergy: a condition in which a person, when stung, may experience a variety of symptoms ranging from a mild rash or itchiness to anaphylactic shock.

Virgin queen: an unmated queen.

Wax glands: the eight glands that secrete beeswax; located in pairs on the last four visible ventral abdominal segments.

Wax moth: larvae of the moths *Galleria mellonella* or *Achroia grisella*, which seriously damage brood and empty combs.

Winter cluster: the arrangement of adult bees within the hive during winter.

Worker bee: a female bee whose reproductive organs are undeveloped.

Worker comb: comb in which workers are reared, and honey and pollen are stored.

FURTHER READING AND ACKNOWLEDGEMENTS

There are many excellent books that can help you to expand your knowledge of beekeeping. Now that you have had a taste of what this fascinating subject is all about and have perhaps taken your first harvest of honey, you may wish to take it further. It is also a good idea to subscribe to a good beekeeping magazine published in your country because it is important to keep up to date with the latest techniques and research, especially where it affects local disease treatments and government compliance legislation.

MAGAZINES
Australia
The Australian Beekeeper
www.theabk.com.au

New Zealand
The New Zealand Beekeeper
https://apinz.org.nz/shop/publications/
the-new-zealand-bee-keeper-journal

Spain
Vida Apicola
www.vidaapicola.com

United Kingdom
BeeCraft
www.bee-craft.com

The Beekeeper's Quarterly
https://beekeepers.peacockmagazines.
com

Bees for Development Journal
www.beesfordevelopment.org

Journal of Apicultural Research
www.tandfonline.com

United States
American Bee Journal
www.americanbeejournal.com

Bee Culture
www.beeculture.com

BOOKS
Atkinson, J., *Background to Bee Breeding* (Northern Bee Books, Hebden Bridge, Yorks, UK, 1999).

Beck, B., *The Bible of Bee Venom Therapy* (Health Resources Press, Silver Spring, Maryland, USA, 1997).

Cook, V., *Queen Rearing Simplified* (Northern Bee Books, Hebden Bridge, Yorks, UK, 2008).

Cramp, David, *The Beekeeper's Field Guide* (How To Books, Oxford, UK, 2011).

Fearnley, James, *Bee Propolis: Natural Healing from the Hive* (Souvenir Press, London, UK)

Hooper, T., *The Bee-Friendly Garden* (Alphabet and Image Ltd, Yeovil, Somerset, UK, 2006).
Kirk, W., *A Colour Guide to the Pollen Loads of the Honey Bee* (IBRA, Cardiff, UK, 2006).

Riches, H., *The Medical Aspects of Beekeeping* (Northern Bee Books, Hebden Bridge, Yorks, UK, 2009).

Van Toor, R., *Producing Royal Jelly,* (Bassdrum Books, Tauranga, New Zealand, 2006).

Wilson, Bee, *The Hive* (John Murray, London, UK, 2004).

Woodward, D., *Queen Bee Biology, Rearing and Breeding* (Northern Bee Books, Hebden Bridge, Yorks, UK, 2009).

ORGANIZATIONS
Hives Save Lives
www.hivessavelives.com

Bees Abroad
www.beesabroad.org.uk

Apimondia (International Federation of Beekeeping Associations)
www.apimondia.com
Apimondia exists to promote scientific, ecological, social and economic apicultural development in all countries and the cooperation of beekeepers` associations, scientific bodies and of individuals involved in apiculture worldwide.

International Bee Research Association (IBRA)
www.ibrabee.org.uk
IBRA is a not-for-profit organization which 'aims to increase awareness of the vital role of bees in the environment and encourages the use of bees as wealth creators'. Its website gives information about the Association's mission,

members, library services and publications, including the Journal of Apicultural Research, and Bee World, a major bee science journal.

Bees for Development
www.beesfordevelopment.org
This is an information service at the centre of an international network of people and organizations involved with apiculture in developing countries. It aims to provide information about beekeeping to alleviate poverty and maintain biodiversity.

Apidologie
www.apidologie.org
This is a major bi-monthly international journal which publishes original research articles and scientific notes concerning bee science and Apoidea.

Centro Andaluz de Apicultura Ecologico (CAAPE)
Based at the University of Cordoba in Spain. It specialises in solutions for organic beekeepers, or those trying to limit their use of chemicals.

Association Nationale des Eleveurs de Reines et des Centres d'Elevages Apicoles (ANERCEA)
The French Association of Queen Rearers.

www.beekeeping.com
The international beekeeping virtual gallery that covers most relevant beekeeping items.

The author and publisher would like to thank the following individuals and companies for their valuable contribution to this book:

John Holmes, Beekeeper
Wildwood Trust, UK
(www.wildwoodtrust.org)

John Laidler
Modern Beekeeping
www.modernbeekeeping.co.uk

Godfrey Munro
Park Beekeeping Supplies
www.parkbeekeeping.com

John Phipps
Editor: The Beekeepers' Quarterly, Greece
webpage: www.iannisphoto.com

Colin Francis Builders Ltd

PICTURE CREDITS

The publisher would like to thank the following photographers and picture libraries for the use of their pictures in the book. Every effort has been made to acknowledge the pictures properly.
Alamy: 14, 15b, 18l, 20b, 29tr, 39r, 41tl, 45br, 49b, 72tr, 74t, 91tl, 91tm, 106, 108 (all), 111bl, 116bm, 116ml, 135tl, 137br, 140t, 145ml, 145tr, 145br, 146t.
Ardea: 20tl, 21tr, 67br, 89tr, 118ml, 131 (row 8).
Bridgeman Picture Library: 12t, 15tr.
Corbis: 94t, 134 (both), 141t, 145bl, 148bl, 150 (all), 153t.
Chris Delker: www.bees-and-beekeeping.com: 50–53 (all except 53l).
Felicity Forster: 4tm, 5tm, 8, 19tm, 31b, 32, 120, 132, 136, 157.
Colin Francis Builders Ltd: 44t.
Getty: 36b, 37t, 101mr, 113tr.
Paul Honigmann of oxnatbees.wordpress.com: 45bl
Istock: 85t, 91tl, 91tm.
John Laidler: 44b, 72bl.
John Phipps: 2, 16, 19b, 42bl, 43bl, 54 (all), 55tl, 55tr, 55m, 59tl, 59tm, 59tr, 65tl, 65tm, 65tr, 67tl, 67tm, 68 (all), 69 (all except mr), 73t, 84br, 90t, 101 (top row), 101ml, 101mr, 103b, 104b, 107br, 109b, 111t, 111bm, 111br, 113b, 115 (all), 118b, 135bl, 144t, 145t, 147tr, 147tl, 148br, 148tr, 149 (all), 152, 153bl, 155.

Nature Picture Library: 12b, 21tl, 29b, 88, 98bl, 110bl, 114, 119b, 131 (row 1, 4, 5, 7, 9).
NHPA: 13b, 37bl, 89bl, 103m, 118t, 119m, 131 (row 3).
Photolibrary: 18r, 19tr, 20tr, 21b, 38bm, 38br, 49tl, 49tr, 58b, 86br, 96, 99b, 144b.
Robert Pickett: 4b, 5tm, 5tr, 5b (both), 6 (both), 7 (both), 10 (both), 11b, 19tl, 22 (both), 23 (both), 24 (all), 25 (all),26 (all), 27 (all), 29tl, 29tm, 34, 35tr, 35br, 36tl, 38t, 38bl, 39l, 40r, 41tr, 41bl, 41bm, 41br, 42tl, 42br, 43t, 46tl, 46bl, 46br, 47 (all), 48 (all), 50tl, 53l, 55bl, 56 (all), 57 (all), 58t, 59b, 62 (all), 63tl, 63tr, 63br, 64 (all) 65 (all), 66 (both), 67bl, 67tr, 69tr, 70, 71 (all), 72tl, 72mr, 73b, 74bl, 74bm, 75t (both), 81 (all), 82, 83 (all), 84bl, 84t, 85l, 86tl, 86tr, 86bl, 87b, 89tl, 89br, 90bl, 90bm, 90br, 91tr, 92 (all), 93 (all), 98tl, 98tr, 98br, 99 (all except 99b), 100 (all), 102r, 103tl, 103mr, 104t, 105 (all), 107tl, 107tr, 107bl, 109tl, 110t, 110br, 112 (all), 113tl, 116bl, 116br, 117 (all), 118mr, 137tl, 137tm, 137tr, 138 (all), 139 (all), 140b, 141b, 143tl, 143tm, 143b, 147br, 148t, 150r, 151tl, 151tr, 154.
SuperStock UK: 13t, 37br, 40l, 43tr, 45t, 63bl, 85b, 95bl, 151b.

INDEX